INTERACTION OF BRIDGE AND VEHICLES
— IN THE BOX-GIRDER —
BRIDGES DESIGN

ALIAKBAR PABARJA

INDIA • SINGAPORE • MALAYSIA

Notion Press

Old No. 38, New No. 6
McNichols Road, Chetpet
Chennai - 600 031

First Published by Notion Press 2019
Copyright © Aliakbar Pabarja 2019
All Rights Reserved.

ISBN 978-1-64678-989-4

Dedication

To my late mother
To the champion of my heart my father

Special Thanks to:

Mr Ata Khorrampoor
Dr Ehsan Jabalbarezi Hookerd
Dr Iman Mohseni
Dr Mohammadreza Vafaei

CONTENTS

INTRODUCTION

Box-girder bridge is employed widely in the development of highway systems and interchanges in urban area to meet man-made obstacles and sever constrains of alignments. Box-girder bridges are introduced to larger torsional and flexural stiffness, which are essential for skewed and curved bridges. Due to the close shape of these bridges, the internal area of boxes is less exposed to environmental damage caused by corrosion. In addition, box-girder bridges supply flat and aesthetically pleasing system as well as economy. There are different types of box-girder bridges such as single box bridges, multicell box-girder bridges, spread box-beam bridges, and multiple-box girder bridges. Concrete box-girder bridges may be built applying prefabricated concrete sections, which are constructed at a manufacturing factory and then transferred to construction site; or utilizing cast-in-place concrete, which is prepared and cast in the final site using false work. To cope with the man-made obstacle, space limitations and complex intersections, the bridge engineers tend to construct skewed superstructures. Despite all benefits of the skewed systems, their general behaviour is more complicated than straight ones. A large number of investigations have been conducted to evaluate the static and dynamic response of box-girder bridges, but there are only a limited number of analytical or experimental studies available on concrete multicell box-girder bridges. The current codes frequently estimate fairly conservative or unsafe value of live load distribution and dynamic responses of this system subjected to vehicle loading conditions. Thus, further study is required to evaluate the effects of live loads on straight and skewed multicell box-girder bridges.

Multicell box-girder bridges are constructed of concrete slabs and webs with either vertical or inclined external webs. Continuous skewed concrete multicell box-girder is three-dimensional and is relatively complicated system. The current bridge design practical specifications have accepted the concept of distribution factor to make easier the evaluation and design of multicell box-girder bridges. However, most live load distribution factor formulas are either; developed for short-single span, straight bridges; or included high complex equations with limited ranges of applicability. In the current bridge design codes concentration were mostly on developing simplified formulas for predicting the distribution of shear force and bending moment due to vehicle loads on bridges, however, the predicting of maximum stress, reaction force and deflection on the multicell box-girder is still a controversial issue. The current bridge design practical specifications have accepted the concept of distribution factor to make easier the evaluation and design of multicell box-girder bridges. However, most live load distribution factor formulas are either; developed for short-single span, straight bridges; or included high complex equations with limited ranges of applicability. In the current bridge design codes concentration were mostly on developing simplified formulas for predicting the distribution of shear force and bending moment due to vehicle loads on bridges, however, the predicting of maximum stress, reaction force and deflection on the multicell box-girder is still a controversial issue. A diaphragm is a lateral stiffener that is set between webs to secure the section geometry. Because of high labour cost of concrete diaphragms, use of intermediate diaphragms (IDs) is considered as an added cost to bridge construction. The advantages of using IDs are still controversial issue among the bridge engineers, and the current bridge specifications do not consider the effectiveness of IDs on live load distribution factors. However, current investigations indicated that configurations and the number of IDs greatly affect lateral and vertical transfer of vehicle load on skewed bridges (Cai et al. 2010; Li & Ma 2010). Thus, further investigation is required to evaluate the effect of IDs on static and dynamic behaviour of multicell box-girder bridges.

In skewed bridges, the torsion moment in the obtuse corner of the bridge and the transverse moments in the deck increase with skew angel (Minalu 2010; Théoret et al. 2012). Torsion includes transverse and negative flexural bending that must be considered in bridge design of skewed bridges. Although the advanced finite element program is able to take these effects into consideration they are often very time-consuming. The grillage analysis requires a post-processing to determine the torsional moment of the superstructure that is partly confusing. Meanwhile, most specifications (AASHTO LRFD 2008 and AASHTO 2002) and simplified methods (Henry's method) are unable to determine the secondary moment of bridges. Thus, further study is required to develop a simply expressions to determine the secondary moment for skewed bridges. There is an inclination in most bridge design codes to deal with static loads keeping away from complex and difficult dynamic analysis. Thus, most bridge is designed by employing static loads which are intensified by the dynamic impact factor (dynamic load allowance). The dynamic impact factor suggested by current specifications are either based on works performed a decades ago on a limited number of actual bridges; or limited to straight bridges with particular cross section. The current investigations indicated that current specification formulas may estimate highly conservative or unsafe values for dynamic impact factor of skewed multicell box-girder bridges (Ashebo et al. 2007a), Therefore an extensive study is required to develop more precise expressions for dynamic impact factor of this type of bridge.

TYPE OF BOX-GIRDER BRIDGES ANALYSES

The determination of the maximum responses, such as stress, shear, reaction, and bending moment for skewed box-girder bridges is a complex task. Until recently, live load distribution factor have been used in bridge design to estimate the effect of vehicle passing over the bridges. The proper prediction of the live load distribution is fully dependent upon the accurate simulation of actual bridges in term of their configurations, boundary conditions, material properties, applied loading system, and also the method of bridge analysis.

The presence of skewness in concrete box-girder bridges caused it to be difficult to precisely estimate their static and dynamic responses to vehicle loading conditions. However, by the use of the advanced computer, the difficulties in the bridge design have been reduced. Since the general behaviour of concrete box-girder bridges subjected to moving vehicle is always elastic, technique of elastic structural analysis like orthotropic and finite strip technique can be employed. However, bridge engineers have also tended to accept formulated and conservative simple methods such as wheel load distribution factor and dynamic load factor to estimate the dynamic and static responses of bridges.

In this chapter, a brief review on the methods of analysis is given. These methods include grillage analogy technique, finite element method, orthotropic plate theory, fload-plate theory, finite strip analysis, and thin-walled-beam theory. In addition, an overview of field test study on the skewed box-girder bridges is also presented. Meanwhile, a brief review of the previous work regarding the live load distribution factor, skew correction factor and dynamic impact factor of skewed box-girder are summarized.

1.1 Feature of Multicell Box-Girder Bridges

Skewed box girder bridges are used extensively in the contribution of highway systems and interchanges in urban areas when severe restrictions of alignments and site conditions exist. Box girder are known to have higher flexural and torsional rigidities, which are required for skewed and curved bridges. Due to their close shape, box girders are less exposed to the environmental detriments causing corrosion. There are different types of skewed box girder bridges. They may be made of reinforced concrete, prestressed concrete, steel, steel box composite with a concrete deck.

Concrete box girder bridges may be constructed using precast concrete elements, which are fabricated at the production plant and then delivered to the construction site, or using cast-in-place concrete, which is formed and cast in its final position using falsework or a launching frame. In the case of prestressed box girder bridges, there are two types of prestressed systems: pre-tensioning and post-tensioning. Pre-tensioning system is method in which the stands are tensioned before the concrete is placed and post-tensioning system are methods in which the tendons are tensioned after concrete has reached a specified strength. Box girder bridges have single or multicell boxes as shown in Figure 2.1.

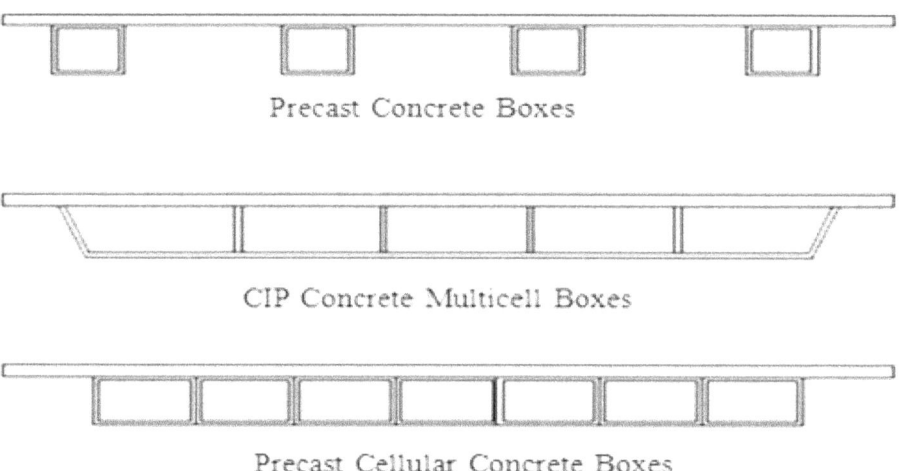

Figure 2.1 Various Types of Box Girder Bridges

1.2   Multicell Box-Girder Bridge Analytical Methods

There are several methods for the analysis of box-girder bridges. In almost all of these methods, the three-dimensional bridge is simulated by applying simple assumptions in the material properties, geometry characteristics, and boundary condition. The preciseness of the bridge analysis widely depends on the selected methods of analysis. Scordelis (1982) has presented a review of the various available bridge analysis methods for concrete box-girder bridges based on several advanced computer program developed at the University of California. The analytical aspect of several bridge analysis methods described by Kristek (1979). Maisel & Roll (1974) performed a comparative investigation on straight single-cell box-girder bridges to evaluate the reliability of various analysis methods. In the following sections, a review of different analytical method for box-girder bridges is demonstrated.

1.2.1   Finite Element Method

The finite element method is being progressively used to structures and is generally the most forceful analysis method employed in the investigation and industry, particularly for the simulating of the behaviour of a box-girder bridge with any arbitrary geometric cross section. In this method the structure is simulated by subdividing the whole bridge into individual elements applying suitable finite elements. The separate element stiffness matrix is arranged to the entire matrix. The nodal displacements and the interior stresses in each element are deduced either by stress method or displacement method. The compatibility and equilibrium of structure are attained by employing adequate mesh refinement (Zienkiewicz & Taylor 2005). Different kinds of elements have been developed for finite element analysis, such as beam elements, shell elements, or solid elements.

Sisodiya et al. (1970) conducted finite element analysis of skew single box-girder bridges with in-plan-curvature. The bridge that has been investigated might have varying width, curved in any shape, not just a circular shape. The rectangular elements were used for the webs and parallelogram or triangular elements for upper and bottom flanges. This estimation would need a large number of elements to attain a reasonable solution. Such an approach is impractical, especially for highly curved box bridges. Chapman et al. (1971) performed a numerical study using finite element technique on steel and concrete box-girder bridges with different cross section shapes to evaluate the influences of internal diaphragms on the warping and distortional stresses. It was concluded that curved steel boxes even with symmetrical load components result in increasing distortion, and revealed that making box-girder bridge with incline external webs leads to an intensify growth distortional stresses.

Lim et al. (1971) expanded a particular element that can be used to analyze straight, skew, or curved box-girder bridges. The new element is trapezoidal in shape including beam-like-in-plane displacement field. Chu & Pinjarkar (1971) developed a finite element mathematical approach for curved box-girder bridges by defining a sector plate element for horizontal direction and cylindrical shell element for vertical components. A sensitive study on cellular bridges with constant depth and various geometric shapes carried out by William and Scordelis (1972). The bridge actions were determined by modeling the prototype bridges with quadrilateral element.

Fam & Turkstra (1975) developed an theoretical finite element method based on the combination of the boundary element method, to investigate the static responses and free vibration analysis of box-girder bridge with orthogonal boundaries and various combination of straight and curved superstructures. The finite-element technique was employed to simulate the webs and bottom flanges of box-girder cross section, while the boundary element method was applied to simulate the bridge deck. The flexural moments and vertical displacement were discovered to be in satisfactory agreement when compared to the finite difference method. Li (1992) developed a box girder finite element, which includes extension, torsion, distorsion, and shear lag analysis of straight, skew, and curved multicell box-girder bridges using thin-walled elements based on Vlasov's theory. Exact shape function were used to eliminate the need for dividing the box into many element in the longitudinal direction. The results of proposed element agreed well with those results obtained from full three-dimensional shell element analysis.

Hodson et al. (2012) developed a finite element modeling method using primarily eight-nodal solid element with three freedom degree at each node to analyze skewed post-tensioned box-girder bridges. The finite element analysis indicates the numerical efficiency, preciseness as well as flexibility in static and dynamic analyses of bridges. Therefore, many researchers are interested in accepting the finite element technique to evaluate the box-girder bridges.

The finite element method has been recommended by both Canadian Highway Bridge Design Code (CHBDC 2000) and The American Association of State Highway and Transportation Officials (AASHTO 2004) to analyze all type of bridges.

1.2.2 Grillage Analogy Method

A grillage analogy method is a technique of describing a bridge superstructure with a series of beam elements that have been allocated properties appropriate for depicting the girder torsional and flexural stiffness, unit weight characteristics of bridge. Grillage models decrease the number of elements and degree of freedom required to analyze, whereby reducing the number of equations that require to be solved for obtain bridge responses.

Hambly & Pennells (1975) and Hambly (1991) applied two types of grillage modeling approaches for straight and skewed box-girder bridges known as ordinary planar and downstand grillage methods. The former method is one in which all longitudinal and transverse members are placed at the location of main members in the bridge deck. A downstand grillage method is a more complicated technique in which the longitudinal and transverse component does not merge into the deck plane. Instead, the upper and lower planes treat as independent grid systems which connected to each other by rigid links. The rigid link secure compatibility condition on the deformation and reactions of bridge superstructure.

Beal & Kissane (1975) developed the grillage methods for the curved multispine box-girder bridges. The curved bridge was simulated as a system of separate curved longitudinal and transverse members interconnected orthogonally. Due to the reduced in stress at points far of webs as results of shear lag, the slab width was substitute with a diminished effective width over which the stress is supposed to be constant. The deflection, flexural bending moment and transverse shear distribution of dead and live load on box-girder bridges were investigated by Cheung et al. (1982) and Burgueño & Pavlich (2008). The finding data indicated good agreement with the corresponding results collected from finite strip analysis and field test study.

One problem regarding the grillage analogy method is in estimating the adequate effective width of bridge slab, to account the shear lag influence. In addition, correct approximation of torsional stiffness for closed boxes is another difficulty related to presented method. Evans & Shanmugam (1984) displayed that simulating the torsional stiffness of a single box-girder bridges using an equivalent I-shaped beams stiffness obtain satisfactory results.

The Canadian Highway design code (CHBDC 2000) has adopted using grillage analogy method for analysis of box-girder bridges in which the number of boxes is more than two. The AASHTO Load and Resistance Factor Design (AASHTO LRFD 2008) performed the grillage method to develop live load distribution formulas for different type of bridges.

1.2.3 *Orthotropic Plate Theory*

Orthotropic plate theory refers to material which has different elastic properties along two orthogonal directions. The method takes into account the relationship between the concrete deck and the idealized beam of a box girder bridge. In this method the rigidity of the intermediate diaphragms is distributed over the girder length and the stiffness of the flanges and girders are massed together into an orthotropic plate of equivalent stiffness. However, the judgment on the bending and torsional stiffness is known as a main problem in orthotropic plate theory. Also, the stress interaction in the slab and girder is another problem in applying this technique. In spite of this, Cheung (1969) developed orthotropic plate theory for determine moment and shear force for curved multispine girder bridges.

Based on several studies on development of orthotropic plate theory, it was revealed that there was good agreement between results obtained using experimental and finite element analysis on box-girder cross section bridges with various shapes (Bakht et al. 1981; Lopez-Anido & GangaRao 1995; Chou et al. 2006).

Higgins et al. (2011) expanded deflection equations for infinitely wide and simply supported thin plates considering each of the three cases of orthotropic: relatively torsionally stiff, flexurally soft; uniformly thick plate; and torsionally soft, flexurally stiff; to evaluated the bridges with various cross section. The developed analytical method allow quick analysis of multiple moving patch loads to obtain maximum live load effects and permit verification of experimental and numerical methods.

The dynamic behaviour of the box-girder bridge under single and several vehicles moving in different lanes was investigated using the orthotropic plate theory by Zhu & Law (2002). However, AASHTO (2002) specifications recommended the orthotropic plat method as a refined approach for bridge analysis, but no guideline is presented. The Canadian highway bridge design (CHBDC 2000) has limited the use of orthotropic plate theory only for evaluating the straight multispine girder bridges.

1.2.4   Folded Plat Method

The folded plate technique, which developed by Scordelis (1960) to determine longitudinal stress, deflection and transverse moment of bridges, provide a reliable solution for linear elastic analysis of simple supported box-girder bridges.

In this method, the bridge simulated as an assembly of longitudinal plate component interconnected at joints along their longitudinal edges and simply-supported at both ends by diaphragms. These diaphragms in their own plane are extremely stiff but in perpendicular direction to their own plane are superbly elastic. In this way, the compatibility and equilibrium requirements are satisfied at the interface elements.

This method can be used for solution of box-girder bridges subjected to any arbitrary load condition which can be presented in the form of harmonic component with use of Fourier series. The folded plate method employs plane-stress theory and classical two-way bending plate to obtain bridge actions. Meyer and Scordelis (1971), Al-Rifaie & Evans (1979) developed the folded plate theory to evaluate the simple-supported cellular bridges. The results of an analytical investigation on simple span multiple box-girder bridges reported by Johnston and Mattock (1967) indicated that folded plate technique can be used to determine the distribution of combined load on girder.

American Association of State Highway and Transportation Officials Load and Resistance Factor Design (AASHTO LRFD 2008) has adopted folded plate theory for analysis all type of bridge cross section. Canadian Highway Bridge design (CHBDC 2000) limited the use of that for bridge with line-support condition.

1.2.5   Finite Strip Method

The finite strip method was resulted from semi-analytical finite element method originally developed by Cheung (1968) for analyzing structures with constant geometrical characteristics along the longitudinal axles. The method subdivides the structure into a number of rectangular strips, running from one end support to another. These rectangular strips interconnected along their longitudinal edges by nodal line. The stiffness matrix of a rectangular strip is determined by minimum potential theorem, based upon a displacement function in term of Fourier series. The direct stiffness harmonic method, as described for folded plate theory, is applied in this technique. The displacement components are developed as a sum of a longitudinal set of functions, which vary from trigonometric series to Eigen-function, and polynomials.

Cheung (1968) and then Cheung & Cheung (1971) developed the finite strip approach to analyze the curved and straight box-girder bridges. Kabir & Scordelis (1974) expanded a computer program based on finite strip method to solve the numerical problem of straight and curved cellular bridges with internal diaphragms. The method was developed to analyze signal and continuous box-girder bridges by Cusens & Loo (1974).

In 1978, Cheung & Chan (1978) reached out a simplified method to calculate the effective width of compression flange of multicell and multispine bridges with use of this method. Cheung (1984) adopted an advanced method established upon the finite strip technique and force method for the evaluation of multi-span curved box-girder bridges. Maleki (1991) and Shimizu & Yoshida (1991) further developed the spline finite strip method for elastic structural analysis of box-girder bridges. Bradford & Wong (1992) employed the finite strip approach with one harmonic to investigate the local buckling of the straight composite box-girder bridges under negative bending. Design graphs are presented to calculate the local buckling coefficient for the web of concrete deck with steel box section bridges.

Using the finite strip technique, Cheung & Foo (1995) evaluated the finding data of a parametric study on the static behaviour of curved and straight box-girder bridges. The parameters assess in the study were; cross section shaped, type and magnitude of applied forces, span length and radius of curvature. Proposed expressions were deduced for flexural moment ratios between curved and straight box-girder structures.

Currently, Cheung & Song (2009) offered a bending crack strip method that joins the shape functions of the spline finite strip with the Eigen-function explanation of the differential expression that govern the deflection around a crack. In this way, the complicated three-dimensional box-girder deck with established cracks that are perpendicular to the longitudinal axis of the spline would be more simplified and can be analyzed sufficiently.

However, the American Association of State Highway and Transportation Officials (AASHTO 2002) has adopted the finite strip method to analyze all type of bridges, the Canadian highway and bridge design (CHBDC 2000) restricted the applicability of this method to simply supported structure with simple line-support.

1.3 Experimental Study on Box-Girder Bridges

Buchanan et al. (1973) performed two sets of experimental studies on a composite twin-box-girder bridge located in the Maryland. The bridge responses were measured before establishing the concrete deck and later, when the structure was completed, and the bridge was under traffic loading conditions. The collected data from field testing was analyzed using finite difference methods. The obtained results indicated a close agreement with the analytical methods. It also showed that bridge design specifications generally overestimate the bridge responses.

Aslam & Godden (1975) conducted an extensive experimental study to examine the accuracy of a computer program provided to analyze the straight, curved, and skewed box-girder bridges. They created a small-scaled aluminum model of a four-cell box girder bridge. The model was experimented elastically, with and without diaphragms, subjected to a concentrated point load. It was concluded that the folded plate theory is adequately precise for determining the elastic responses box cross section bridges.

In 1975, Kissane & Beal (1975) conducted a field test program of a horizontally curved, two-span, composite bridge with three-spine boxes, located in New York. The bridge actions contain strain, deflection; rotation and deformation of cross-section were measured. The finding results were compared with a analytical solution of bridges which developed based on planar grid and stiffness method to obtain various bridge responses at each joint. It was concluded that experimental bending moment and deflection for dead and live load compared well with their corresponding analytical responses. However, due to high torsional stiffness caused by intermediate diaphragms, the lower live load distribution factor compared to those from current specifications was obtained. Similar Field study was conducted by Yoo et al. (1976) on a continuous three-span, Twin-spine curved box-section bridge located in Baltimore.

In 1985, Shanmugam & Balendra (1985) employed the finite element method and experimental study to evaluate the elastic behaviour of two Perspex multicell box-girder bridges. Two types of webs were simulated-one with solid web and another with web openings. Results from analytical method compared reasonably well with the experimental finding for stress and deflections.

Siddiqui & Ng (1988) and Arizumi et al. (1988) tested elastically several simple supported straight and curved box-girder bridges, to investigate the effect of transverse diaphragms in decreasing the distortional and warping stresses. The rectangular and trapezoidal cross-section was considered in this study. The finding results indicated that deformations of box-girder may results in considerable warping and distortional stress. They also concluded that by setting up sufficient number of rigid diaphragms along the span, the warping and distortional stresses can be restrained. The similar results were obtained by Park et al. (2003) that developed a box beam element having nine degrees of freedom per node including two distortional degrees of freedom to obtain the desired ratio of the distortional warping normal stress to the bending normal stress. They also proposed tentative design charts for adequate maximum spacing of intermediate diaphragms to cope with distortional warping in horizontally curved steel box girder bridges (Park et al. 2005).

Okeil & El-Tawil (2004) performed a case study on 18 composite steel-concrete box-girder bridges to consider the effect of warping on normal bridge responses. Finding results indicated that warping has little effect on both shear and normal stress in all bridges. Ebeido & Kennedy (1996) conducted an experimental study on three continues skewed composite steel box, concrete deck with unequal span length. The bridge was tested up to collapse using simulated vehicle load passing over to load lane.

The load distribution and dynamic responses of composite cellular bridges were investigated by Sennah (1998). Five straight and curved prototype bridges were selected and tested subjected to different load conditions, and free vibration analysis was performed to determine the mode shapes. The similar study were conducted to aid in better understanding of static and dynamic responses of straight and curved multiple box-girder bridges (Samaan et al. 2002; Samaan 2004; Samaan et al. 2007). Based on the finding data, empirical equations were derived to determine the load distribution factor and dynamic impact factor of straight and curved bridges. Furthermore, Moon et al. (2005) tested the stress variation and deformation properties of a box-girder segment during the construction of actual bridges. A construction pattern for preventing crack during construction was also developed.

Dynamic behaviour of long-span bridges with box-girder shapes, tested by Lee & Yhim (2005). A four node Lagrangian and Hermite finite elements were used to numerically analyze of bridges under passing vehicle over the superstructure. The experimental data from a real two-span prestressed concrete single-box bridge was used to verify the numerical analysis.

Richardson & Douglas (1993) conducted a field testing on a curved box girder highway overpass. In the study, large horizontal loads were applied to the deck of the bridge and quickly liberated, causing the bridge to vibrate. The resulting high-amplitude vibrations were destined to resemble the seismic load caused by earthquakes. Well-defined lateral, longitudinal, vertical and torsional vibration modes were identified from the test data.

Richardson & Douglas (1993) used vibration modes to verify an analytical model of the bridge's dynamic response. For that study, the model was versified using only the fundamental vibration mode, which was primarily a horizontal vibration mode. Using a system identification procedure, the dynamic response model was adjusted until its frequency and mode shape matched the measured frequency and mode shape. Parameters in the verified model were compared with the same parameters calculated from information in the structural drawings. Because the fundamental mode represents a horizontal mode, the bridge parameters identified in this paper were those parameters which strongly influence the horizontal response of the bridge.

Ashebo et al. (2007 a, b) reported on field testing study of an existing multicell box-girder bridge, located in Hong-Kong, to assess the skew effect on dynamic bridge-vehicle responses. The influence of skewness on the static and dynamic behaviours of the bridge as well as on the load distribution in the transversal direction for the calibration truck and in-service vehicles was investigated. It was found that the influence of skew in both the static and dynamic behaviours of the bridge within the skew angle range of 0–30° is very small. It was also observed that the well-planned experimental on bridges can get wide ranges of high-quality values to determine the dynamic responses of bridges.

Zhang et al. (2010) investigated the effect of shear lag in streamlined thin-walled box-girder bridges with large width-to-span ratios through both experimental and analytical methods. A large-scale Plexiglas model is tested under various loading condition. It was indicated that the finite element method is an effective technique to determine responses of this bridges.

Huang et al. (2004) and Huang (2008) conducted a full-scale test on curved steel-concrete box-girder bridges. The experimental results indicated: (1) current AASHTO guide specifications regarding the first transverse stiffener spacing at the simple end support of a curved girder may be too conservative for bridge load capacity ratings; (2) current AASHTO guide specifications may greatly overestimate the dynamic loadings of curved box girder bridges with long span lengths; and (3) a plane grid finite-element model of about 20 elements per span in the longitudinal direction can be used to analyze curved bridges with external bracings located only over supports.

LIVE LOAD DISTRIBUTION
FACTOR OF BOX-GIRDER BRIDGES

For many decades, the live load distribution factor has been known as a relatively simple method to predict the maximum responses of bridge, such as bending moment, shear, and deflection. With the simplified equations for distribution factor and beam-line methods, the wheel load distributed on each member can be assigned. The live load distribution on bridges depends on the location/type and magnitude of applied loads, and varies with changing in bridge characteristics, such as number of boxes, skewness, number of lane loads.

Until recently, the load distribution factor for straight and skewed bridges was simply determined by the expression S/7 and S/5.5 for single and multilane loaded bridges, respectively. In these expressions, S is the girder spacing and no influences of skewness and curvature were taken into account.

Since 1967, Henry's simplified EDF method (TDOT 1996) has been used by Tennessee Department of Transportation to estimate the wheel load distribution factor of all type of bridges. The Henry's method assumed that both external and internal girder received the same portion of applied loads. Because Henry's method requires only the width of the roadway, number of traffic lanes, number of beam lines, and the intensity factor of the bridge, it can be applied without difficulty to different types of superstructures and beam arrangements. For most bridges, the distribution factors obtained from Henry's method are smaller than the ones from the AASHTO Standard Specifications (Huo et al. 2005).

Twenty-four Tennessee bridges of six different types of superstructures were selected for detailed analysis and comparison by Huo et al. (2003). Finite element analysis was pursued to determine the moment and shear distribution factors for each of these bridges. Based on the comparison and evaluation, it was found that in the case of straight bridges, the Henry's distribution factors was in a good agreement with the moment distribution factors obtained from FEA and the LRFD method and were consistently unconservative for shear distribution factors compared to the FEA results.

Johnston & Mattock (1967) expended a computer program to estimate the maximum live load distribution in simple supported, box-girder bridges. Mattock & Fountain (1968) implemented the folded plate method in a computer program to calculate the lateral distribution of loads in 24 simply supported composite multiple box girder bridges. The results of the folded plate computer program were verified by testing one-quarter model of a two-lane, 24.4 m span bridge with three box girders and one-fifth model of a two-lane, 30.4 m span bridge with two box girders, under AASHTO truck loadings. The results obtained from the computer program were used to develop an expression for the live load bending moment distribution factor for each girder as a function of the roadway width and the number of box girders. The results from the research program by Fountain and Mattock formed the basis for the lateral distribution factors for bending moment currently adopted by AASHTO (1996) and Ontario Highway Bridge Design Code (1983) for multiple box girder bridges. The results were obtained based on ratio of relatively limited investigation of a number of bridges considering only the number of lanes and number of boxes as variables. Most importantly, the curvature and the continuity effects were not considered in the study.

Heins (1978) surveyed and collected the geometric properties of 82 actual bridges with span length ranging from 50 to 250 ft. built until 1975. The finding data were used to simulated a typical composite concrete-steel model of box-girder Bridge. The prototype model was analyzed to provide a modified factor to account for curvature.

The National Highway Research Program (NCHRP) Project 12-26 (Zokaie et al. 1993) was conducted to assign simplified and practical formulations for live load distribution of truck loads passing over the box-girder bridges. The project defined three level of analysis technique to determine the maximum responses to truck loading condition, including simplified load distribution equation, grillage analysis, and finite element approach. The more reliable equations extended in NCHRP 12-26 Project have been adopted by AASHTO Load and Resistance Factor Design (LRFD) specifications (AASHTO LRFD 2008), and were replaced those specified in the AASHTO standard codes. The recent edition of AASHTO Load and Resistance Factor Design (LRFD) specifications (AASHTO LRFD 2008) contains extended modifications to the initial distribution factor, to consider the effect of skewness and continuity. Table 2.1 and Table 2.2 listed the moment and shear distribution factor equations from AASHTO LRFD specifications for various box-girder bridges.

When the line supports are skewed and the difference between skew angles of two adjacent lines of Supports do not exceed 10°, the bending moment in the beams may be reduced and shear at the obtuse corner may be increased. To consider the skew effect, The AASHTO LRFD specification suggested the following skew correction factor equations for the maximum bending moment at midspan and shear at obtuse corner in accordance with Table 2.3.

Several investigation were conducted on distribution of shear force and reaction on skewed box-girder bridges (Ebeido & Kennedy 1996). Based on finding from experimental and analytical methods, the empirical formulas were suggested for shear distribution factor under designated OHBDC truck loading condition and dead load. The results indicated that the more the bridge is skewed, the more the AASHTO standard specifications overdue the maximum bridge responses. Hence, it was found that AASHTO standard specification fail to reliably define actual behaviour of bridges, such as maximum bending moment at midspan and shear and reaction at the obtuse corner.

To evaluate the live load distribution factors of curved and straight composite open box-girder bridges with various geometries, Samaan & Sennah (2002) and Samaan (2004) conducted extensive experimental and numerical studies. The finding data from a parametric study were used to develop the empirical equations to determine the maximum longitudinal stress, deflection, reaction and shear subjected to dead and traffic loading condition. However the proposed equations were assessed with only one design instance, thus validity cannot be confirmed. More importantly, the effect of skewness did not taken into consideration in developing proposed equations.

Table 2.1 AASHTO LRFD Formulas for Distribution of Live Loads per Lane for Moment (g)

Type of Superstructure	Live Load Distribution Factor		Range of Applicability
	Interior girder	Exterior girder	
Cast-In-Place concrete multicell box	- one design lane loaded $$\left[1.75+\frac{S}{1100}\right]\left[\frac{300}{L}\right]^{0.35}\left[\frac{1}{N_B}\right]^{0.45}$$ - two or more design lane loaded $$\left[\frac{13}{N_B}\right]^{0.30}\left[\frac{S}{430}\right]\left[\frac{1}{L}\right]^{0.25}$$	$\dfrac{W_e}{4300}$ $\dfrac{W_e}{4300}$	$2100 \le S \le 4000$ $18000 \le L \le 73000$ $N_B \ge 4$ IF $N_B > 8$ USE $N_B = 8$ $W_e \le S$
Concrete Deck on Concrete Spread Box Beam	- one design lane loaded $$\left[\frac{S}{910}\right]^{0.35}\left[\frac{S_d}{L^2}\right]^{0.25}$$ - two or more design lane loaded $$\left[\frac{S}{1900}\right]^{0.6}\left[\frac{S_d}{L^2}\right]^{0.125}$$ If S > 5500 use Lever Rule Method	IF $0 \le d_e \le 1400$ Use $g = e \cdot g_{interior}$ $e = 0.97 + \dfrac{d_e}{8700}$	$1800 \le S \le 5500$ $6000 \le L \le 43000$ $450 \le d \le 1700$ $N_B \ge 3$
Concrete Deck on Multiple Steel Box Girder	Regardless of number of loaded lanes $$0.05 + 0.85\frac{N_L}{N_B} + \frac{0.425}{N_L}$$		$0.50 \le \dfrac{N_L}{N_B} \le 1.5$

(Source: AASHTO LRFD. 2008)
L: Span Length (m); N_B = Number of boxes; N_L = Number of lanes; d= depth of bridge; S=web spacing; W_a= width of Bridge; S_c= box spacing

Table 2.2 AASHTO LRFD Formulas for Distribution of Live Loads per Lane for Shear

Type of superstructure	Distribution factor		Range of applicability
	Interior girder	Exterior girder	
Cast-In-Place concrete multicell box	- one design lane loaded $$\left[\frac{S}{2900}\right]^{0.6}\left[\frac{d}{L}\right]^{0.1}$$ - two or more design lane loaded $$\left[\frac{S}{2200}\right]^{0.9}\left[\frac{d}{L}\right]^{0.1}$$	- one design lane loaded Lever Rule - two or more design lane loaded $g = e \cdot g_{interior}$ $e = 0.64 + \dfrac{d_e}{3800}$	$1800 \le S \le 4000$ $6000 \le L \le 73000$ $890 \le d \le 2800$ $N_C \ge 3$ $-600 \le d_e \le 1500$
Concrete Deck on Concrete Spread Box Beam	- one design lane loaded $$\left[\frac{S}{3050}\right]^{0.6}\left[\frac{d}{L}\right]^{0.1}$$ - two or more design lane loaded $$\left[\frac{S}{2250}\right]^{0.8}\left[\frac{d}{L}\right]^{0.1}$$ Use Lever Rule	$g = e \cdot g_{interior}$ $e = 0.8 - \dfrac{d_e}{3050}$ $0 \le d_e \le 1400$ Use Lever Rule	$1800 \le S \le 5500$ $6000 \le L \le 43000$ $450 \le d \le 1700$ $N_b \ge 3$ S > 5500
Concrete Box Beam Used in Multibeam Deck	- one design lane loaded $$0.70\left[\frac{b}{L}\right]^{0.15}\left[\frac{I}{J}\right]^{0.05}$$ - two or more design lane loaded $$\left[\frac{b}{4000}\right]^{0.4}\left[\frac{b}{L}\right]^{0.1}\left[\frac{I}{J}\right]^{0.05}\left[\frac{b}{1200}\right]$$ $\dfrac{b}{1200} \ge 1.0$	- one design lane loaded $g = e \cdot g_{interior}\left[\dfrac{1200}{b}\right]$ $\dfrac{1200}{b} \le 1.0$ $e = 1 - \left[\dfrac{d_e - b - 610}{12200}\right]^{0.5} \ge 1$	$900 \le b \le 1500$ $6000 \le L \le 37000$ $5 \le N_b \le 20$ $1.0 \cdot 10^{10} \le J \le 2.5 \cdot 10^{10}$ $1.7 \times 10^{10} \le I \le 2.5 \cdot 10^{10}$

(Source: AASHTO LRFD. 2008)
L: Span Length; d= Bridge Depth; S=Girder Spacing; I= Moment of Inertia; J= Torsional Stiffness; d_e= Overhang Length; b=girder spacing

Table 2.3 Skew Correction Factor (SCF) from AASHTO LRFD Specification

Type of Superstructure	SCF for Moment	SCF for Shear	
	Equation	Equation	Range of Applicability
Cast-In-Place concrete multicell box		$1.0 + \left[0.25 + \dfrac{12.0\,L}{70\,d}\right]\tan\theta$	$0 \leq \theta \leq 60^0$ $1100 \leq S \leq 4900$ $6000 \leq L \leq 73000$ $N_b \geq 4$
Concrete Deck on Concrete Spread Box Beam	$1.05 - 0.25\tan\theta \leq 1.0$ If $\theta > 60^0$, then use $\theta = 60^0$	$1.0 + \left[0.25 + \dfrac{L}{70d}\right]\tan\theta$	$0 \leq \theta \leq 60°$ $1800 \leq S \leq 4000$ $6000 \leq L \leq 73000$ $900 \leq d \leq 2700$ $N_b \geq 3$
Concrete Box Beam Used in Multibeam Deck		$1.0 + \dfrac{\sqrt{Ld}}{6S}\tan\theta$	$0 \leq \theta \leq 60°$ $1800 \leq S \leq 3500$ $6000 \leq L \leq 43000$ $450 \leq d \leq 1700$ $N_b \geq 3$
Concrete Deck on Multiple Steel Box Girder		$1.0 + \dfrac{L\sqrt{\tan\theta}}{90d}$	$0 \leq \theta \leq 60°$ $6000 \leq L \leq 37000$ $430 \leq d \leq 1500$ $900 \leq b \leq 1500$ $5.0 \leq N_b \leq 20$

(Source: AASHTO LRFD, 2008)

L: Span Length (m); N_B= Number of boxes; N_L= Number of lanes; d= depth of bridge; S=web spacing; W_a= width of Bridge; S_d= box spacing; θ= skew angle

Zhang (2008) developed the new equations for the skew correction factor (SCF) to eliminate the limitations on the range of applicability and improve the accuracy of AASHTO LRFD specification. Two set of SCFs equations were derived, one set was based on the current AASHTO LRFD specifications, and another established on NCHRP 12-26 Projected (Zokaie et al. 1993). In addition, the similar equations for skew correction factor (SCFs) for reaction distributions of vehicle loads were also derived to account for the distinction between reaction and shear in the skewed bridges. The accuracy of equations were validated by comparison data from finite element and those from proposed equations, however, the equations were highly complicated.

Zheng (2009) provided an insight into the design of skewed and curved composite box girder bridges. Based on the rigorous analysis carried out on 640 prototype steel open-box girder bridges, it is discovered that the span length, number of lane loaded, number of boxes, girder width, skew angle and radius of curvature are the most critical parameters that affect the load distribution factors. It was showed that the distribution factor of external and internal girder are different. Thus, the external girder and internal girder were acted distinctly in the equation development. The extended equations are feasible to straight, skewed and curved bridges. The finding from the study also supplied an insight into the design of skewed and curved composite box-girder bridges.

The analytical and theoretical studies were performed by some researchers (Kozhikote 1989; Li 1992; Chun 2010; Li & Genmiao Chen 2011) to evaluate the effect of skewed supports on distribution of live loads on bridge superstructure. A simplified live load distribution factor framework was given by (Hodson et al. 2011). The effects of analysis ambiguity, changeability, and multiple presences were set apart and obviously explained. This separation obtained specification writers with the occasion to utilize various multiple-presence and models. Live load distribution factor were obtained by employing several simplified techniques and grillage methods for over 1500 prototype bridges. Calibration factors were applied to enhance the preciseness of data.

The influence of parapet, vehicle position, overhang, intermediate diaphragms and aspect ratio on live load distribution factor was investigated by many researchers. Patrick et al. (Patrick et al. 2006) and Huo & Zhang (Huo & Zhang 2008) concluded that the distribution factors are relatively insensitive to vehicle spacing. Therefore significant computational speedups are available when applying vehicle loads on an influence surface with a fixed spacing. Conner & Huo (2006) observed that the presence of parapets reduced distribution factors by as much as 36% and 13% for exterior and interior girders, respectively. The effect of parapets is slightly less for skewed bridges. Aspect ratio is shown to have very little effect on distribution factors until the ratio exceeds 1.8.

Dicleli and Erhan (2010) developed live load distribution formulas for the girders of single-span integral abutment bridges (IABs). For this purpose, two and three dimensional finite-element models (FEMs) of several IABs are built and analyzed. In the analyses, the effects of various superstructure properties such as span length, number of design lanes, prestressed concrete girder size, and spacing as well as slab thickness are considered. The results from the analyses of two and three dimensional FEMs are then used to calculate the live load distribution factors (LLDFs) for the girders of IABs as a function of the above mentioned parameters. The LLDFs for the girders are also calculated using the AASHTO formulas developed for simply supported bridges (SSBs). The comparison of the analyses results revealed that LLDFs for girder moments and exterior girder shear of IABs are generally smaller than those calculated for SSBs using AASHTO formulas especially for short spans. However, AASHTO LLDFs for interior girder shear are found to be in good agreement with those obtained for IABs.

Fanous et al. (2011) developed simple relationships for calculating live-load distribution factors for glued-laminated timber girder bridges with glued-laminated timber deck panels. The live-load distribution factors obtained from the field test and the analytical models were compared with those obtained using the AASHTO LRFD Bridge Design Specifications live-load distribution relations. The comparison showed that the live-load distribution factors obtained by using the AASHTO LRFD Bridge Design Specifications were conservative. For this reason, statistical methods were used to develop accurate relationships that can be used to calculate the live-load distribution factors in the design of glued-laminated girder bridges.

Razaqpur et al. (2013) preformed Extensive measurements during the test to allow better understanding of the response of slab-on-girder bridges as well as their live load distribution characteristics at all stages of loading up to failure. The experimentally determined distribution factors for the tested bridge model are compared with the calculated values based on the Canadian Highway Bridge Design Standard, and the code values are found to overestimate the maximum moment in the interior loaded girder by about 22% and 33% at the elastic and the inelastic states, respectively.

Kim et al. (2013) studied on the flexural behaviour and corresponding load rating of simply-supported steel bridges subjected to military truck loads. The military trucks are categorized by the Military Load Classification (MLC) system according to the North Atlantic Treaty Organization (NATO). The response of the bridges under the MLC trucks is compared with that under the standard HS20 trucks of the American Association of State Highway and Transportation Officials Load and Resistance Factor Design (AASHTO LRFD) specifications. Existing predictive models for load distribution factors are evaluated, including the applicability of bridge code provisions for the MLC trucks. The load rating methods based on the Load Factor Rating (LFR) and the Load and Resistance Factor Rating (LRFR) are studied.

2.1   Studies on Effectiveness of Intermediate Diaphragms

Foinquinos et al. (1997) evaluated the effect of cross frames on distribution factor in straight steel multiple box-girder bridges, within the context of AASHTO standard specifications. It was concluded that setting up only two cross frames improve the stress distribution of live load, and providing more number of cross frames does not have significant effect on results.

Park et al. (2005) extended an extensive study on steel box-girder bridges to develop a curved box beam finite element and to suggest tentative design flowchart for adequate maximum spacing of intermediate diaphragms.

The verification of the proposed box beam element having nine degrees of freedom per node comprising two distortional degrees of freedom was validated from a series of complete comparative investigation using conventional shell element technique.

Previous studies on intermediate diaphragms (IDs) in box-girder bridges often concentrated on whether intermediate diaphragms raised or reduced the live load distribution factors, but no important effort was carried out to extend a formula to quantify the intermediate diaphragm effect on load distributions. Current bridge codes (CHBDC 2000; AASHTO 2002; AASHTO LRFD 2008) understand the need to use intermediate diaphragms for skewed box-girder bridges and also express that intermediate diaphragms affect the vehicle load distribution in bridges. However, during the procedure of growing formulation for live load distribution factors, their effects were not taken into account. The current AASHTO (2002) specifications do not considered the effects of intermediate diaphragms on load distributions, though they adopt those intermediate diaphragms influences.

Yang et al. (2010) conducted Quasi-static and explicit dynamic numerical finite-element analyses of prestressed concrete girder bridges with intermediate diaphragms, and key factors (that is, location and size of intermediate diaphragm, spacing and types of girders, and dynamic load types) involving the role of intermediate diaphragms in the impact protection were evaluated. The bridge without intermediate diaphragms was not capable of sustaining the full design load of 530 kN, demonstrating the important role of intermediate diaphragms on impact protection and performance enhancement of the bridges under impact. It was also observed that for a long bridge, multiple and distributed intermediate diaphragms resisted impact better by effectively transferring large deformations to other girders and the deck, reducing the damaged areas and absorbing more kinetic energy.

Yang & Fu (1997) and then Fu & Tang (2001) developed a computer program based on soften truss modeling technique to torsional analysis of box-girder bridges, however, the effect of skewness on distribution of torsion were not considered.

Minalu (2010) developed an appropriate finite element modelling technique, which capable of predicting the three-dimensional behaviour of high skew bridges consisting of a cast in-place concrete deck on precast prestressed inverted T-girders. Five different numerical models have been created and compared using SCIA engineer and ATENA 3D finite element packages. It was found that the model consisting of shell elements for the deck and eccentric beam elements for the girders is the best for engineering practice. The results show that, live load maximum bending moments in girders of skew bridges are generally smaller than those in right bridges of the same span and deck width. On the contrary, the torsion moment in the obtuse corner of the bridge and the transverse moments in the deck increase with skew angel.

2.2 Dynamic Impact Factor

The estimation of dynamic responses of box-girder bridges is of feasible importance in the design of the bridge. As a result of tendency toward the construction of high-strength material in the past few decades, the bridge members have changed to slender shapes which have more complicated behaviour than traditional bridges subjected to dynamic load due to earthquake, passing vehicle and wind.

In many cases, the critical bridge actions subjected to moving traffic condition across the deck may exceed those determined regarding only the equivalent static loads. Traditionally, to consider the effect of dynamic bridge-vehicle interaction, the static loads are increased by a dynamic load allowance (DLA) factor or dynamic impact factor (DIF). Extensive efforts have been conducted to evaluate the dynamic behaviour of curved and straight bridges since 1970s; comparatively very little attempts have been reported to assess the dynamic behaviour of skewed bridges.

In 1972, Shore & Chaudhuri (1972) have reported the analytical study on I-beam bridge in evaluating the dynamic and static responses of superstructure subjected to moving load. The tentative dynamic impact factor expressions were deduced for stress, reaction and deflection of straight and curved I-beam bridges.

Razaqpur & Li (1994) investigated the dynamic responses of curved box-girder bridges, using finite element model of nine simple supported bridges. The new equations were provided for dynamic impact factor, which later adopted by American Association for State Highway and Transportation Officials, Guide Specification for Horizontal Highway Bridge (AASHTO 1993).

In 1981, Heins & Lee (1981) conducted a field test study on a two-span continues curved box-girder bridge in Korea. The findings were used to better insight into dynamic bridge-vehicle interaction. The results of a field testing study on 27 real bridges, with different configurations and with span length varying from 5 to 122 meter used by Billing (1984) to develop the dynamic load allowance expressions. The proposed expressions adopted by second edition of Ontario Highway Bridge Design Code (OHBDC 1983) and Canadian Standard Association (CSA 1988).

The Ontario Highway Bridge Design Code (OHBDC, 1983) and Australia's National Road Authority (AUSTROADS 1992) recommended a conservative dynamic load allowance factor formula that was only a function of the first flexural frequency of the superstructure. These codes specified dynamic load allowance (DLA) equal to 0.2, 0.4 and 0.25, for bridge superstructure and main longitudinal girder with a first fundamental frequency less than 1.0 Hz, between 2.5 and 4.5 Hz, and higher than 6.0 Hz, respectively, with a linear transition between the range 1.0–2.5 and 4.5–6.0 Hz. Unfortunately, this values is useable only for bridges where the flexural mode is dominant in the longitudinal direction.

Many codes including AASHTO (2002) standard specification (2004) Japanese Road Association (JRA 1996), Canadian Highway Bridge Design Code (CHBDC 2000), Iranian Specification for Highway Bridge Design (Tahouni 2003), and French Cahier des Prescriptions Communes (1973), West German Code; DIN 1072–1976, which referenced by O'Connor and Shaw (2000), developed formula for dynamic impact factor as function of span length of bridge. The recommended formulas for dynamic impact factor from various bridge codes tabulated in Table 2.4.

Table 2.4 Codified Dynamic Load Allowance Formulas

Bridge Code	Dynamic Load Allowance (DLA)
AASHTO Standard (2003)	$DLA = \dfrac{15.4}{L(m)+38.10} \leq 30\%$
Japanese Road Association (1996)	$DLA = \dfrac{20}{L(m)+50}$ For concrete bridges
West German Code; DIN 1072-1976	$DLA = 0.4 - 0.008L(m) - 0.1h \quad 0 \leq DLA \leq 0.30$
Iranian Specification (2003)	$DLA = 0.3 - 0.005L(m) - 0.15 \quad h \geq 0$
French Cahier Descriptions Communes (1973)	$DLA = \dfrac{0.64}{0.20L(m)+1}$ For concrete bridges $DLA = \dfrac{0.80}{0.20L(m)+1}$ For composite bridges

Many researchers (Kashif & Humar 1990; Kashif 1992; Chang & Lee 1994; Akoussah et al. 1997) recommended the use of finite element technique to obtain the dynamic effect of wheel load on box-girder bridges. Liao & Lin (1995) and Sennah et al. (2004) conducted the extensive numerical study to evaluate dynamic bridge-vehicle interaction, and deduced empirical expressions for the dynamic impact factor (or dynamic amplification factor) for moment, shear, reaction and deflection of box-girder bridges.

Zhang et al. (2003) performed a numerical study on 120 concrete-steel cellular bridges to determine the effects of different variable on dynamic resposes of bridges. Based on the data generated from the parametric study, it was concluded that the currents specifications obtain conservative values for dynamic impact factor of bending moment and estimate unsafe values for dynamic impact factor of reaction. Therefore, the expressions for dynamic impact factors for moment, reaction, and deflection for such bridges are proposed. The experimental study were performed by Brady & O'Brien (2006) and Brady et al. (2006) to investigate the effect of vehicle velocity on dynamic responses of simple supported bridges. A set of critical velocities were determined associated with peaks of dynamic amplification for all beams. The reasons for these large amplifications are discussed. A more complicated finite element model, validated with field tests, is used to test the applicability of the conclusions obtained from the simple models to a realistic bridge/vehicle system. Senthilvasan et al. (2002) showed that for box-girder bridges, the dynamic amplification factor in strain are larger than those in deflection. Ashebo et al. (2007 a, b) presented a consideration of vehicle-induced dynamic loads, based on a field test that was performed on an actual skewed multicell box-girder bridge located in Hong-Kong. The influence of various parameters such as number of axles, longitudinal and transverse position of trucks, and vehicle speed were evaluated. The finding results indicated a weak relationship between the speed and dynamic impact factor of bridges. Based on the statistical analysis, the use of the dynamic load factor (DLF) was recommended. The dynamic load factor obtained was less than the values provided by most bridge design specifications.

In contrast, Moghimi & Ronagh (2008) concluded that the dynamic bridge response is significantly influenced by the vehicle velocity. Dynamic load allowance is vehicle dependent and the dynamic and static live loads can be considered uncorrelated, except when the truck weight is less than 10 percent of the total deck weight, for which a low degree of correlation is observed. The DLA is decreased as the vehicle lane eccentricity (with respect to the deck centerline) is increased, and the same relationship exists with the bridge span length. The effects of various parameters on dynamic responses of railway steel bridges and also impact factor values are studied by Hamidi and Danshjoo (2010). Dynamic analysis results show that in most cases the calculated impact factor values are higher than that recommended by the relevant codes and so the offered rations for impact factor are underestimated. It has also been shown that the train velocity affects the impact factor, so that the value of impact factor rises incredibly with the train velocity. Another effective element for impact factor is the ratio of train axle distance to bridge span length so that the impact factor value varies for the ratio below and above unity. The train number of axles just affects the impact factor under resonance conditions. In this paper some relations are offered for the impact factor considering parameters: velocity, train axle distance and the bridge span length. As a results of previous researches, skewed continues concrete multicell box-girder bridges achieved very little attention to assess their dynamic behaviours. Impact factor recommended by current specifications were developed based on inspection only limited number of non-skewed bridges that often estimate very conservative value for skewed bridge. Therefore, a sensitive evaluation needed to improve the accuracy of dynamic impact factor of bridge specification with considering the effect of skewness and continuity in box-girder bridges.

From investigations presented in the prior sections, it seems that most recommended live load distribution formulas are either developed for short-single, straight box-girder bridges; or included high complicated formulas with limited range of applicability. It also concluded that the current practical bridge codes suggested live load distribution factors only to determine the maximum shear and moment, but the distribution of maximum deflection, stress and reaction forces on bridge section is still almost unknown. The finite element analysis is the most reliable method for analysis of box-girder bridges, however the complexity of modeling may suffer researcher. Thus, it is a necessity to obtain reliable equation to determine the live load distribution factor and dynamic impact factor of multicell box-girder bridges. According to previous study, it was found that the current AASHTO specification does not take into account the effect of intermediate diaphragms on live load distribution factor of bridges. The current investigation showed that configurations and number of diaphragms has a great most reliable impact of bridge responses. It was observed that the dynamic impact factor (DIF) or dynamic load allowance (DLA) recommended by AASHTO specifications predict highly conservative or unsafe value for skewed box-girder bridges. However, many investigations have been conducted to evaluate the dynamic bridge-vehicle interaction, only a limited number concentrated on skew box-girder bridges. Thus, a sensitive study on dynamic behaviour of multicell box-girder bridge is required.

2.3   Live Load Distribution Factor for Tensile Stress

In the following, the effect of affecting parameters on distribution factor of tensile stress was investigated.

2.3.1   Effect of Span Length

The results for the bridge span length on tensile stress (positive stress) distribution factor for three and six box-girder bridges were presented in Figure 2.1 to Figure 2.2.

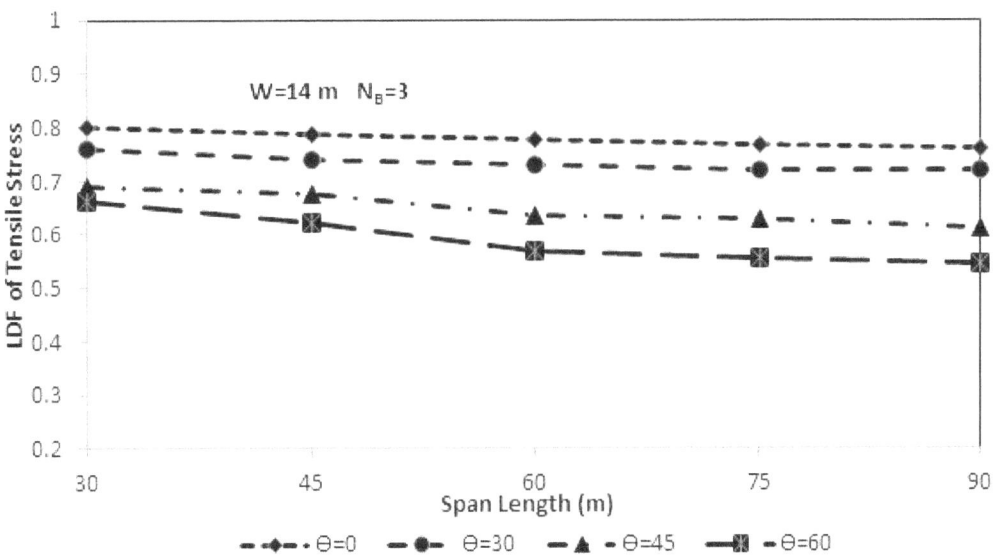

Figure 2.1 Tensile Stress Distribution Factor of vs. Span Length for Three Box Bridges

It can be observed that the live load distribution factor of tensile stress decreased, when span length increased. For example, the live load distribution factor for straight three-box bridges varied from 0.799 to 0.760 (by almost 5%), when span length increased from from 30 m to 90 m. However, the decreased in live load distribution factor for bridge with higher skew angle was more significant, so that for bridge with skew angle of 60°, the distribution factor decreased from 0.662 to 0.544 (decreased by up to 21%) for three-box girder bridges. In same trend can be observed from six-box bridges, so that for bridge with skew angle 60°, the live load distribution factor for tensile stress decreased from 0.335 to 0.290 (decreased by almost 15%), when the span length increased from 30 m to 90 m. Thus, the span length of bridge was considered as a main variable in developing the proposed equation for live load distribution factor of tensile stress.

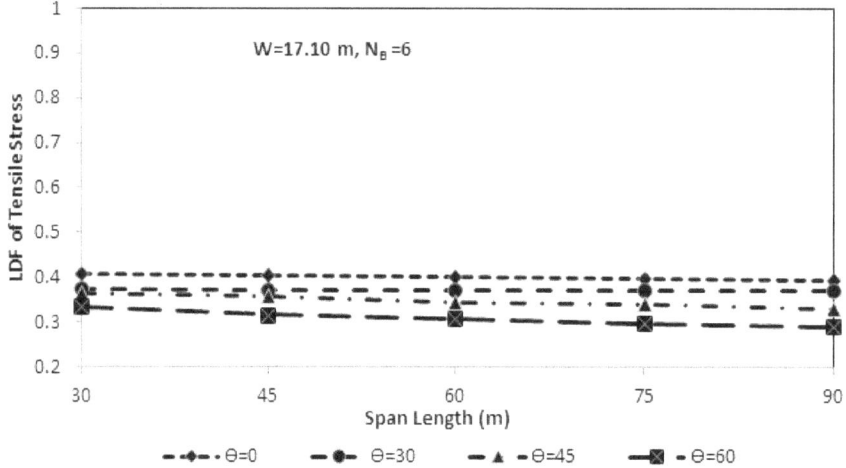

Figure 2.2 Tensile Stress Distribution Factor of vs. Span Length for Six-Box Bridges

2.3.2 Effect of Number of Lanes

The number of lanes represented the bridge width, was evaluated to be one of the critical parameters affecting the lateral load distribution factor of tensile stress on skewed and straight bridges.

Figure 2.3 and Figure 2.4 show the influence of changing in the number of lanes on tensile stress distribution factor of a multicell box-girder bridge with three box-girders, and skew angle of 0 and 45°. The effects of skew angle on tensile stress distribution factor are considered in section 6.2.4.

It can be seen that number of lanes has an increasing effect on maximum tensile stress distribution factor of bridges. For instance, for a prototype bridge with span length of 30 m, by increasing the number of lanes from two to four, the distribution factor of tensile stress varied from 0.610 to 0.830 (increased by 26.5%) for straight bridges and changed from 0.485 to 0.710 (increased by 32%) for bridges with skew angle of 45°. In general, the tensile stress distribution factor at midspan would be lower with increase the span length of bridges.

Thus, the number of lane loads of bridge was considered as a main variable in developing the proposed equation for live load distribution factor of tensile stress.

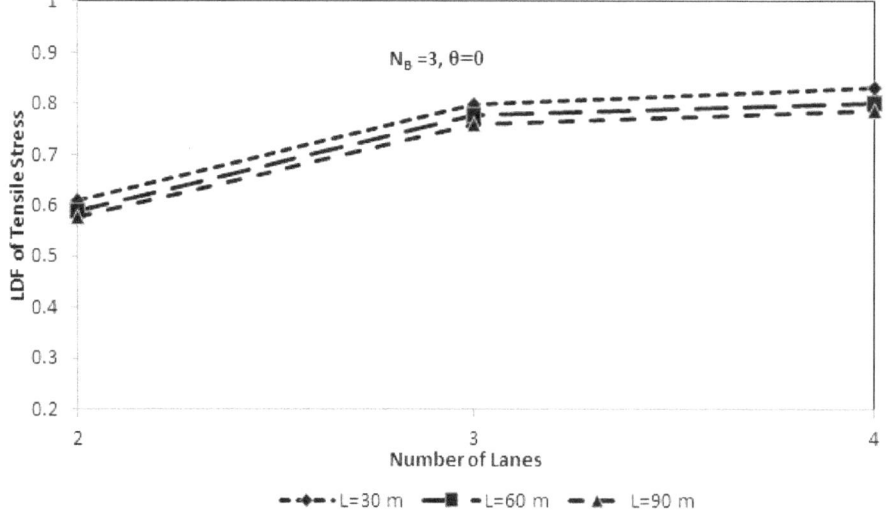

Figure 2.3 Tensile Stress Distribution Factor of vs. Number of Lanes for Straight Bridges

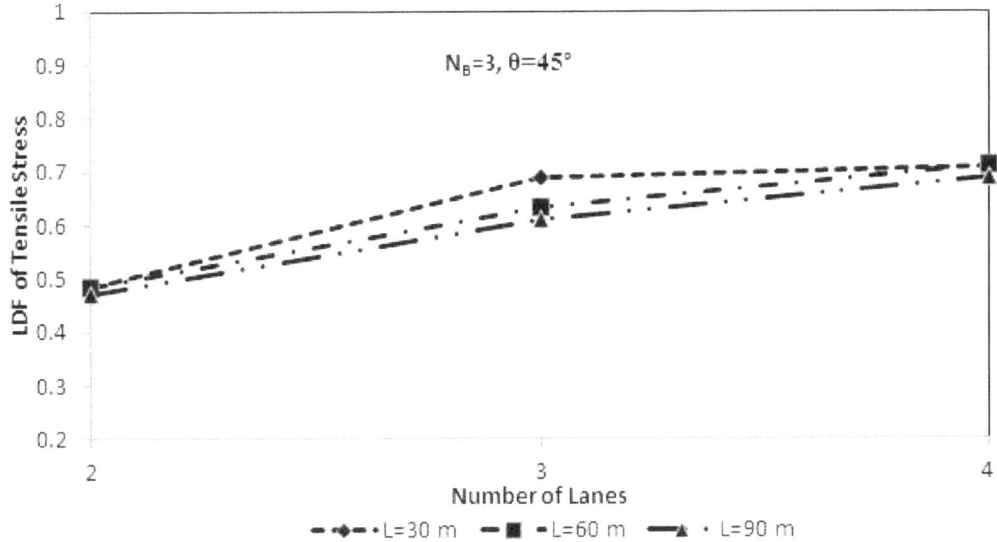

Figure 2.4 Tensile Stress Distribution Factor of vs. Number of Lanes for Bridge with Skew Angle 45°

2.3.3 Effect of Number of Boxes

The effect of number of boxes on the tensile stress (positive stress) distribution factor for multicell box-girder bridges with four lanes, skew angle of 0 and 45°, and having various span lengths were indicated in Figure 2.5 and Figure 2.6.

It can be observed that tensile stress distribution factor decreased from 0.820 to 0.407 (decreased by almost 51%) as the number of boxes increased from three to six, for bridge with span length 30 m. For bridges with skew angle of 45° and with span length of 30 m, the tensile stress distribution factor decreased from 0.71 to 0.364 (almost 48%) when number of boxes increased from three to six. It also concluded that the effect of span length was negligible in variation of number of boxes. The above finding revealed that the increase in the number of boxes may results in a cost effective design in most cases in term of tensile stress.

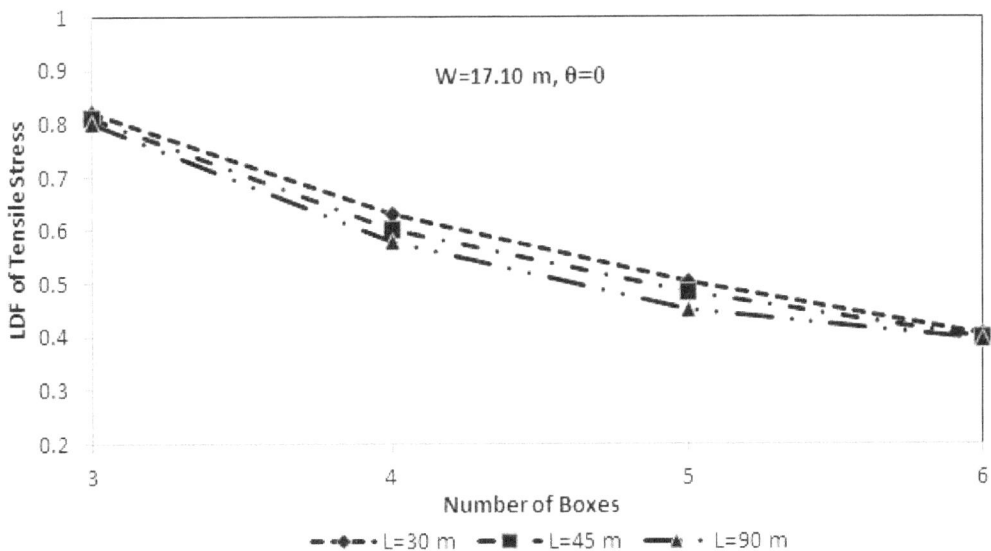

Figure 2.5 Tensile Stress Distribution Factor of vs. Number of Boxes for Straight Bridges

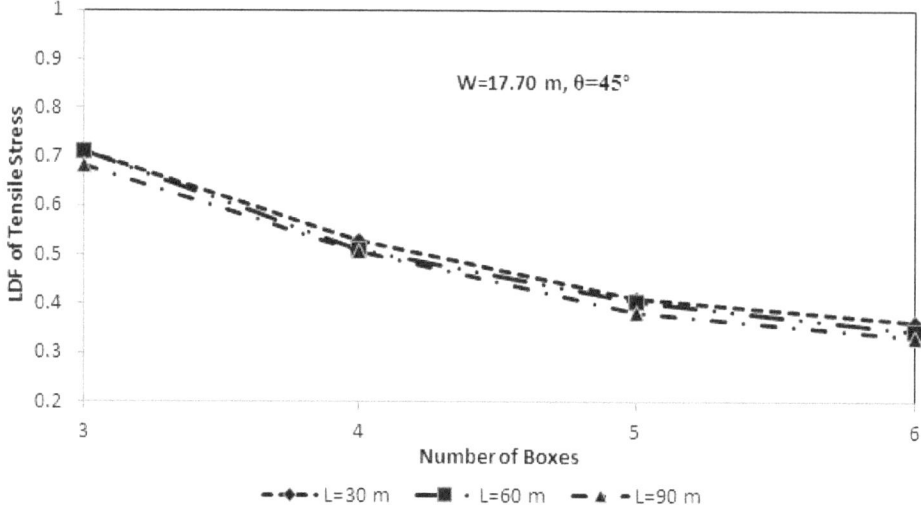

Figure 2.6 Tensile Stress Distribution Factor of vs. Number of Boxes for Bridge with Skew Angle 45°

2.3.4   Effect of Skew Angle

The relationships between the live load distribution factors for tensile stress versus skew angle for three and four lane loaded bridges were plotted in Figure 2.7 and Figure 2.8, respectively. It can be observed that the distribution factors for live load distribution of positive stress decreased when the skew angle increased.

For both three and four lane loaded bridges, when the skew angle is less than 30°, the distribution factor for tensile stress slightly decreased when skew angle increased by almost 5% and 6.5%, respectively. For instant, when the skew angle exceeded from 30°, the distribution factor decreased rapidly from 0.780 to 0.588 by almost 24% for three box bridges with 90 m span length. Tensile stress distribution factor also decreased from 0.57 to 0.404 for four box bridges.

Based on the results, it can be concluded that the skew angle has a significant effect on tensile stress distribution factor of bridges. Thus, the effect of skew angle should be considered in developing new equation for distribution factor of tensile stress at mid-span of bridges.

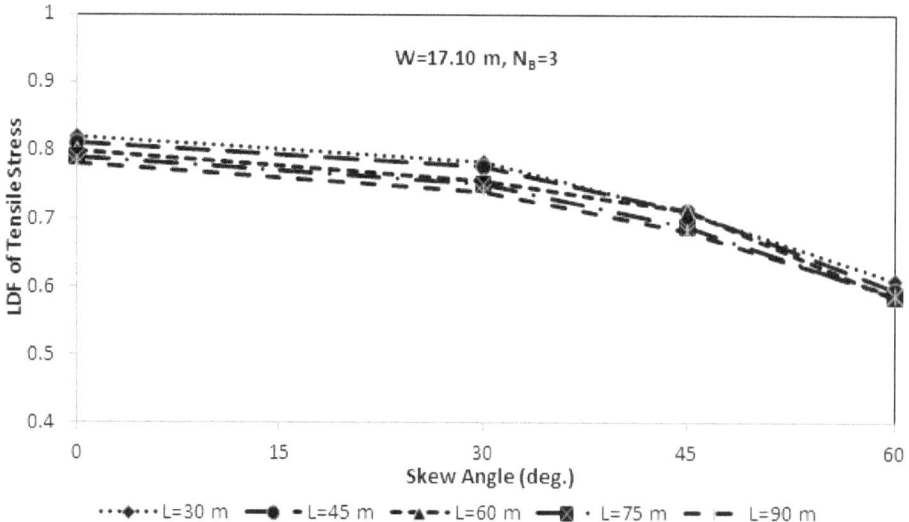

Figure 2.7 Live Load Distribution Factor of Tensile Stress vs. skew angle for Four Lane Loaded Bridges

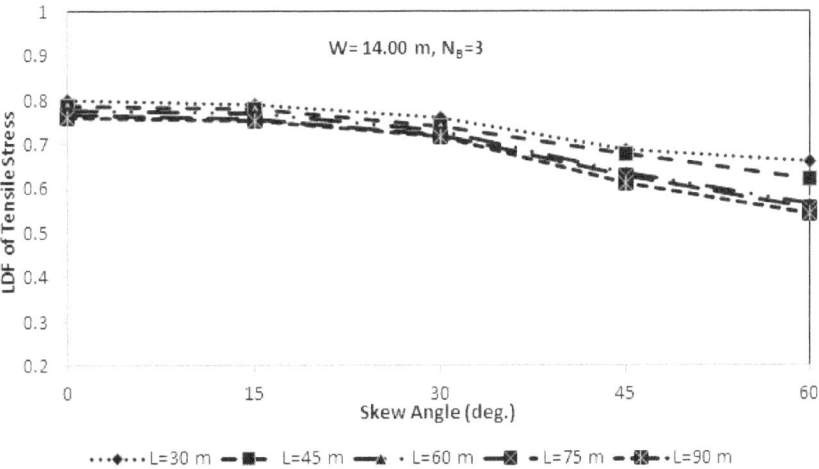

Figure 2.8 Live Load Distribution Factor of Tensile Stress vs. skew angle for Three Lane Loaded Bridges

2.4 Comparison of Analytical Results with Current Specifications

Analytical results of tensile stress distribution factors of bridges with span length of 30 m, 60 m, and 90 m with total bridge width of 17.10 m, were compared with LRFD formulas and AASHTO standards specifications, as indicated in Figure 2.9 through Figure 2.11.

It can be seen that the distribution factor of tensile stress obtained from finite element analysis were significantly smaller than that determined through LRFD formulas and AASHTO (2002) standards specifications. For example, for a bridge with span length of 30 m and 60 m and skew angle of 45°, the difference between tensile stress distribution factor obtained from finite element method and AASHTO LRFD (2008) were up to 20% and 34%, respectively. The difference between LDF for tensile stress obtained from FEA and AASHTO (2002) specifications were almost 40% for bridges with span length of 30m and 60 m. It is due to the fact that AASHTO LRFD (2008) specifications recommended about the use of corresponding bending moment distribution factor to compute the maximum stress on bridge superstructure.

Based on the finding results, it was concluded that the current United States bridge codes were too conservative in term of tensile stress distribution factor of multicell box-girder bridges. Overestimating of tensile stress distribution factor may results in uneconomic design of bridges. Therefore, it was essential to update the live load distribution factor formulas for tensile stress of this type of bridges.

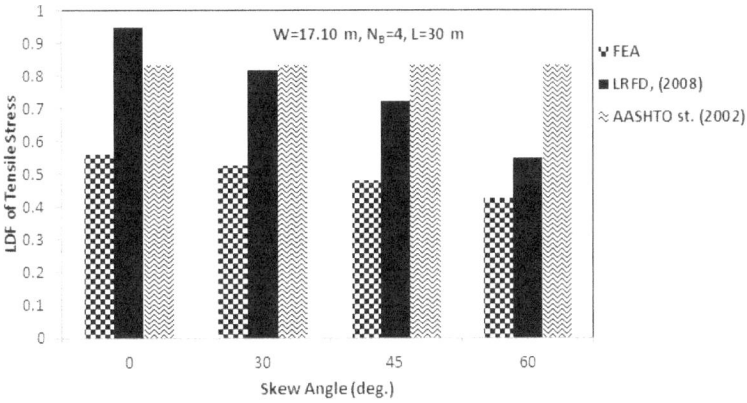

Figure 2.9 Live Load Distribution Factor of Tensile Stress from Various Analytical Methods for 30 m Bridges

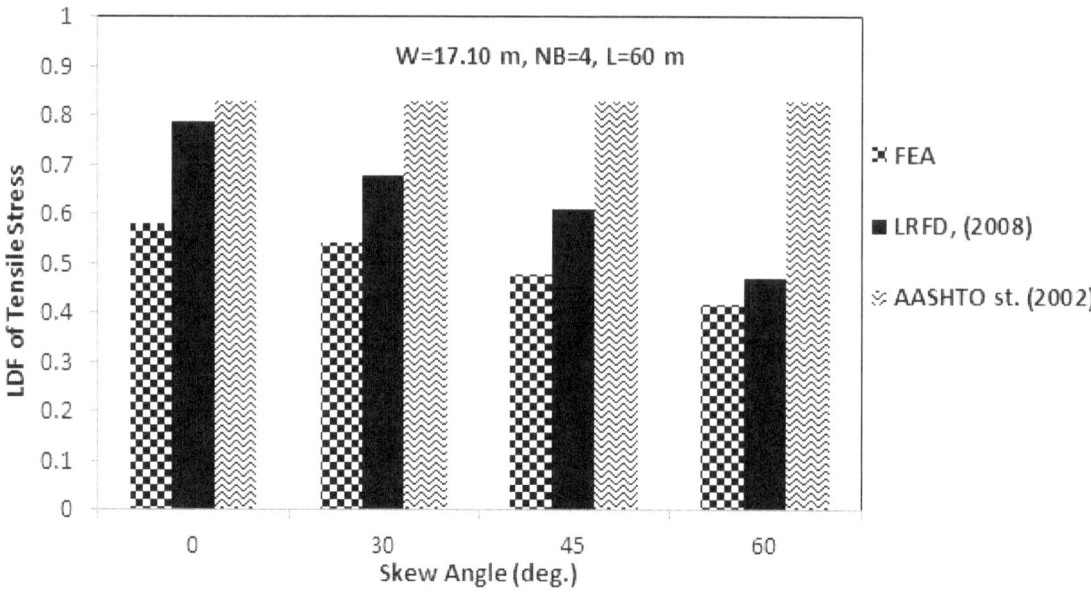

Figure 2.10 Live Load Distribution Factor of Tensile Stress from Various Analytical Methods for 60 m Bridges

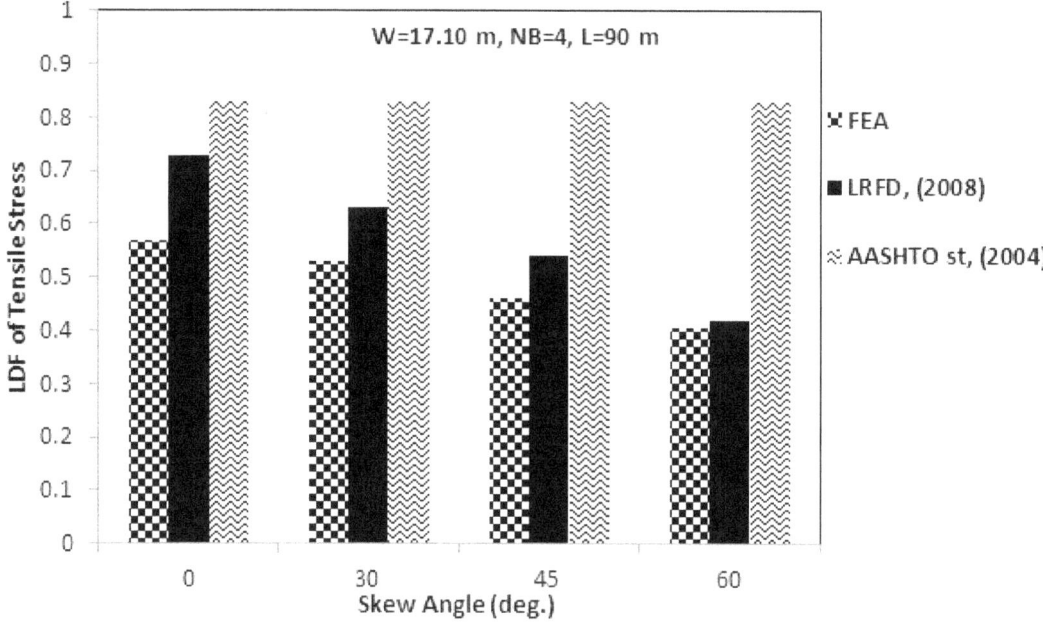

Figure 2.11 Live Load Distribution Factor of Tensile Stress from Various Analytical Methods for 90 m Bridges

2.4.1 Live Load Distribution Factor for Compressive Stress

The effect of main parameters on live load distribution factor of compressive stress (negative stress) was considered in the following sections.

2.4.1.1 Effect of span length

The results for the bridge span length effect on compressive (negative) stress distribution factor for three, four and six box-girder bridges are indicated in Figure 2.12 to Figure 2.14.

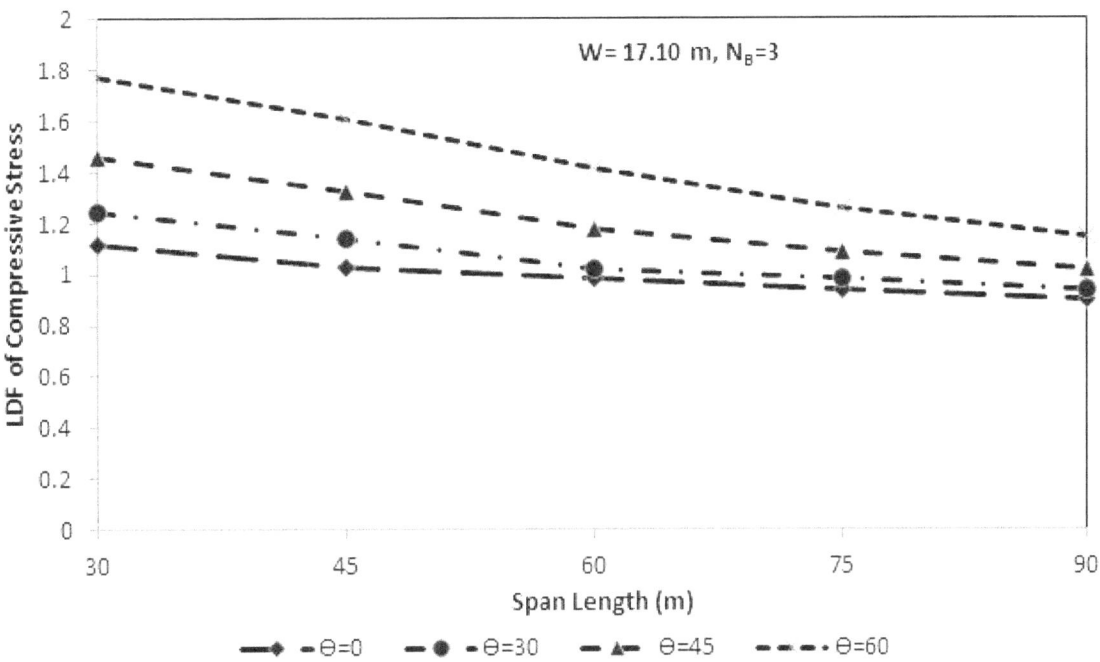

Figure 2.12 Compressive Stress Distribution Factor vs. Span Length for Three Boxes Bridge

It can be observed that the maximum compressive stress (negative stress) distribution factor which obtained at the piers of continuous bridges, decreased with increase of the span length. The finding results indicated that variations of distribution factor of compressive stress was more significant for bridge with greater skew angle, so that the compressive stress distribution factor for a four box-girder bridge changed almost 20% and 32% for skew angle of 0 and 60°, respectively, with increasing the span length from 30 m to 90 m. Thus, the effect of span length cannot be ignored, when developing the equations of distribution factor for compressive distribution factor of bridges.

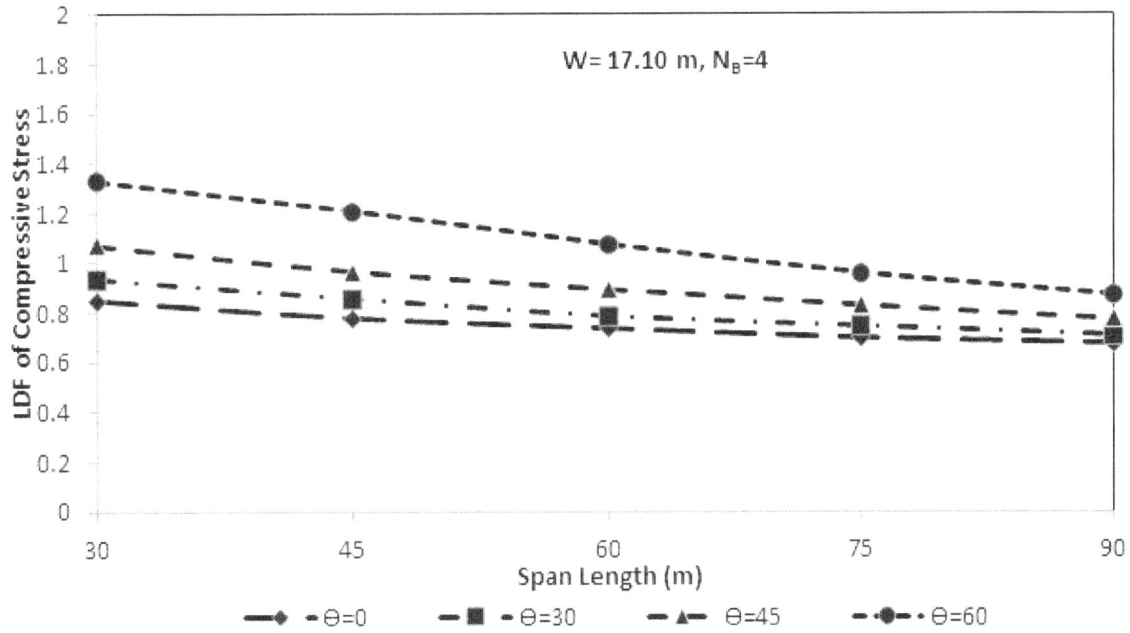

Figure 2.13 Compressive Stress Distribution Factor vs. Span Length for Four-Box Bridges

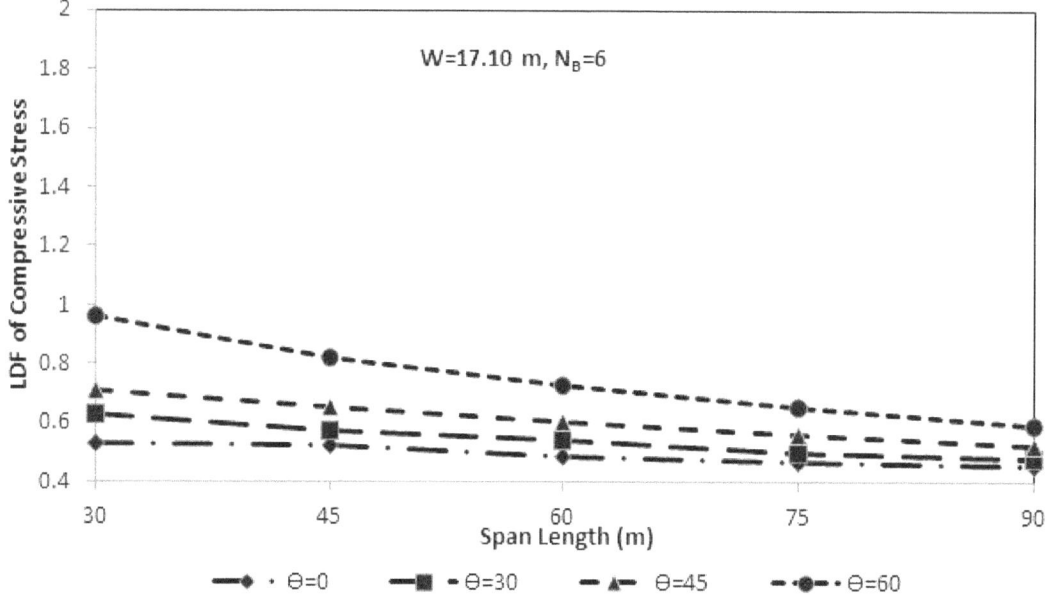

Figure 2.14 Compressive Stress Distribution Factor vs. Span Length for six-Box Bridges

2.4.1.2 Effect of number of lanes

The relationship between the distribution factors for compressive stress versus number of boxes for two, three and four lanes bridges, with three boxes and skew angle of 0 and 60°, are presented in Figure 2.15 and Figure 2.16.

It was observed that the number of lanes has an increasing influence on maximum compressive distribution factor of bridges. When number of lanes increased from two to four, the distribution factor increased by average 33% for different span length of bridges.

It also was concluded that the compressive stress distribution factor with various span lengths followed nearly the same trends, when number of lane loading increased. Thus, the number of lanes should be considered as critical parameters in developing the new equation for live load distribution factor of compressive stress for multicell box-girder bridges.

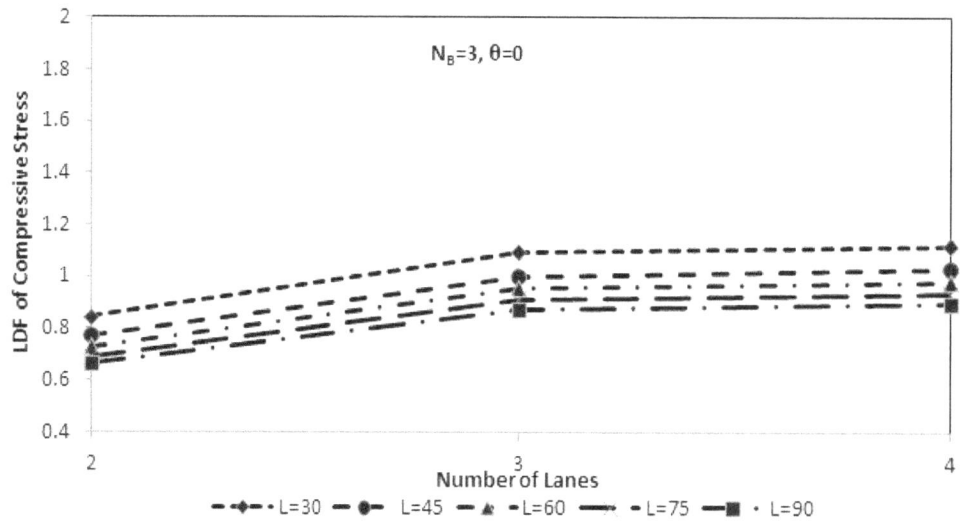

Figure 2.15 Compressive Stress Distribution Factor vs. Number of Lanes for Straight Bridges

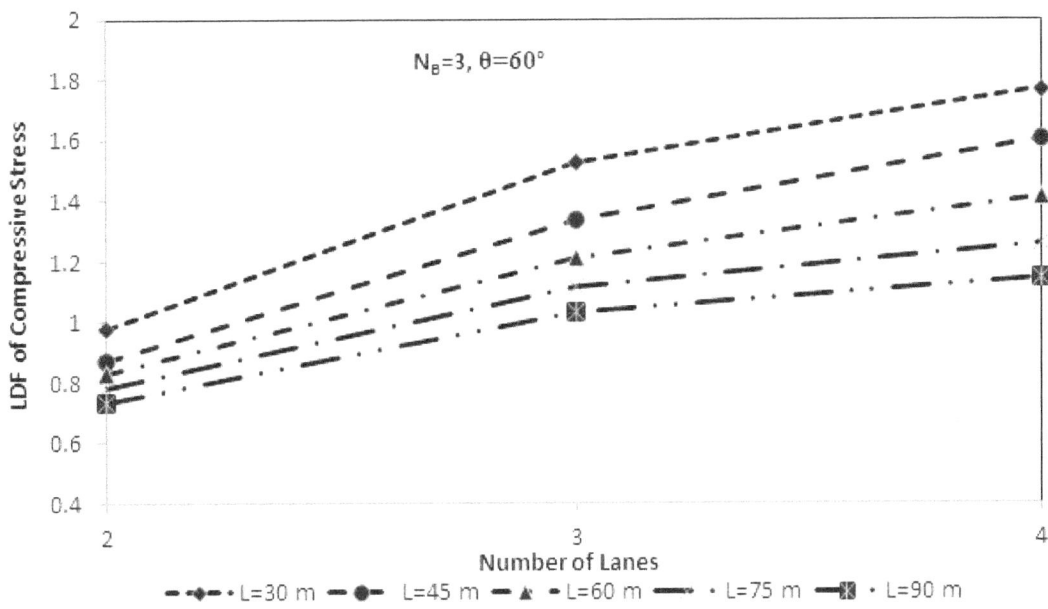

Figure 2.16 Compressive Stress Distribution Factor vs. Number of Lanes for Bridges with Skew Angle of 60°

2.4.1.3 Effect of number of boxes

The relationship between the number of boxes, representing the girder spacing, and the distribution factors for maximum compressive stress (negative stress) at piers for bridge with three and four lanes loading, are shown in Figure 2.17 and Figure 2.18.

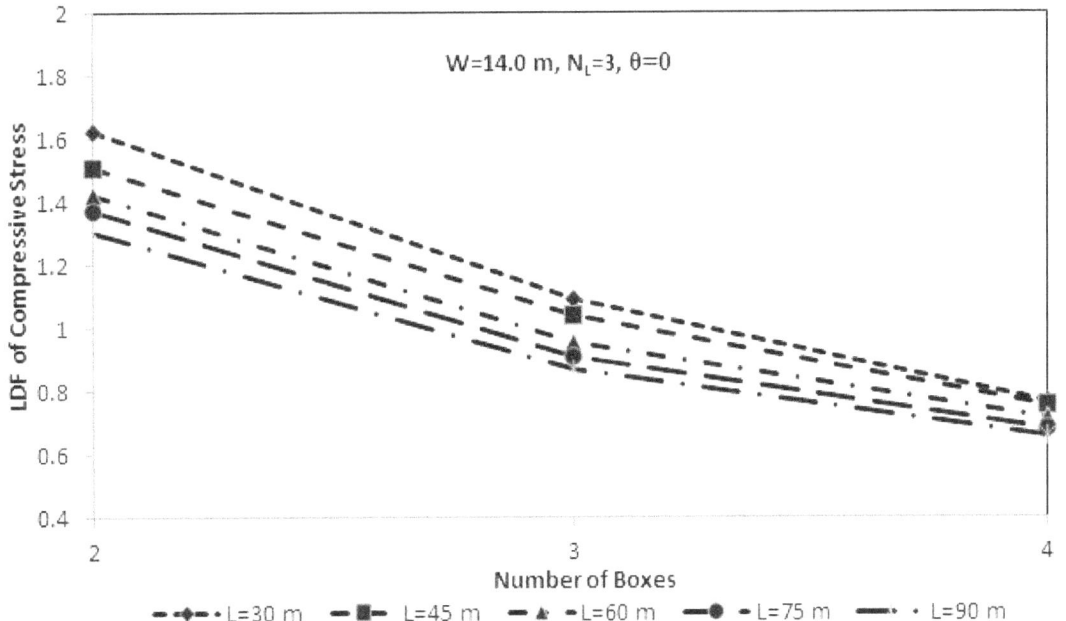

Figure 2.17 Compressive Stress Distribution Factor vs. Number of Boxes for Straight Bridges

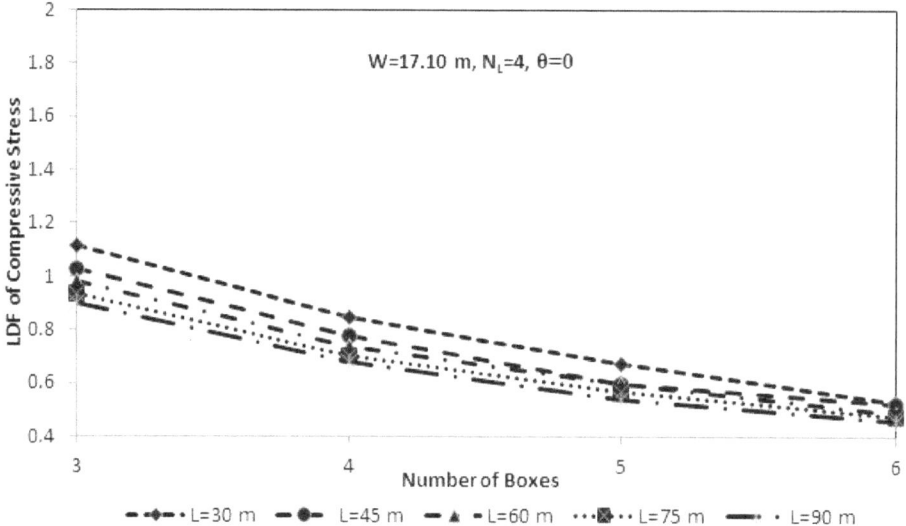

Figure 2.18 Compressive Stress Distribution Factor vs. Number of Boxes for Four Lane Loads Straight Bridges

It can be concluded that the live load distribution factors decreased by almost 52%, when the number of boxes increased from two to three. The variation of distribution factor versus the number of boxes followed nearly the same trend for different span length. Therefore, the effect of number of boxes should be taken into account in developing equations for distribution factor of compressive stress subjected to truck loading conditions.

2.4.1.4 Effect of skew angle

The relationships between the distribution factors for live compressive stress versus skew angle for three and four box-girder bridges, with span length ranged from 30 m to 90 m, are plotted in Figure 2.19 through Figure 2.20, respectively.

It can be observed that the distribution factors for compressive stress increase with an increase of skew angle. For bridges with span length of 30 m, when the skew angle is less than 30°, the average increase of compressive stress distribution factor was about 13.5% and 5.2% for bridge with larger span length, respectively. With skew angle increased from 30° to 60°, the average increase of compressive stress distribution factor was about 52.4% and 130.2% for bridge with span length of 30 m and 75 m, respectively.

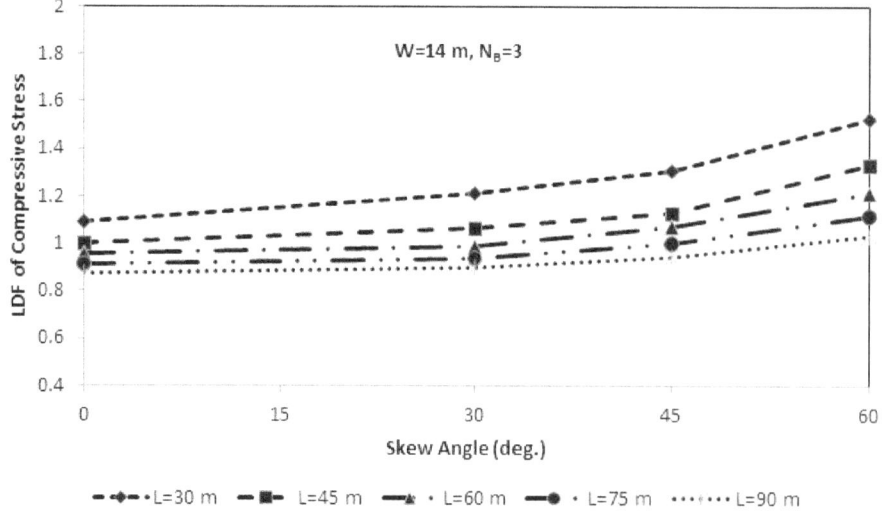

Figure 2.19 Compressive Stress Distribution Factor vs. Skew Angle for Three Lane Loaded Bridges

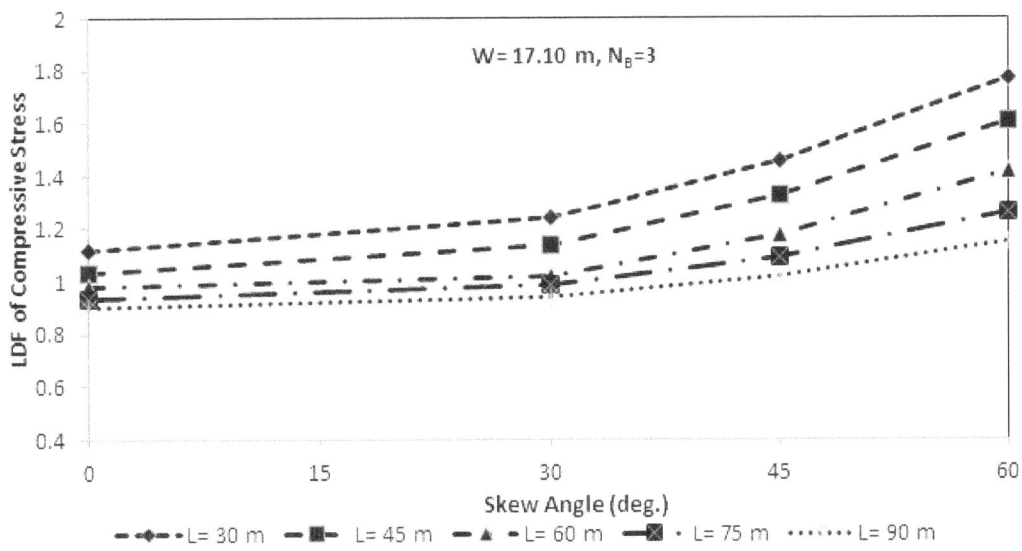

Figure 2.20 Compressive Stress Distribution Factor vs. skew angle for Four Lane Loaded Bridges

It also can be observed that the compressive load distribution factor with different number of lane loaded followed the similar trend. The finding results indicated the neglecting the effect of skew angle on distribution factors may results in unsafe design of bridges in term of compressive stress. Thus, the skew angle considered as a critical parameters in developing the equation for live load distribution factor of compressive stress.

2.4.1.5 *Comparison of analytical results with current specifications*

As described in before, the AASHTO LRFD (2008) specifications and AASHTO (2002) standard code did not proposed any equations to determine the maximum stress distribution factor of bridges.

Traditionally, the stress distribution factor of bridges is assumed to be equal to corresponding moment distribution factor. Analytical results of distribution factor for compressive stress of bridge with span length of 30 m, 60 m and 90 m, with skew angle ranged from 0 to 60° shown in Figure 2.21 through Figure 2.23.

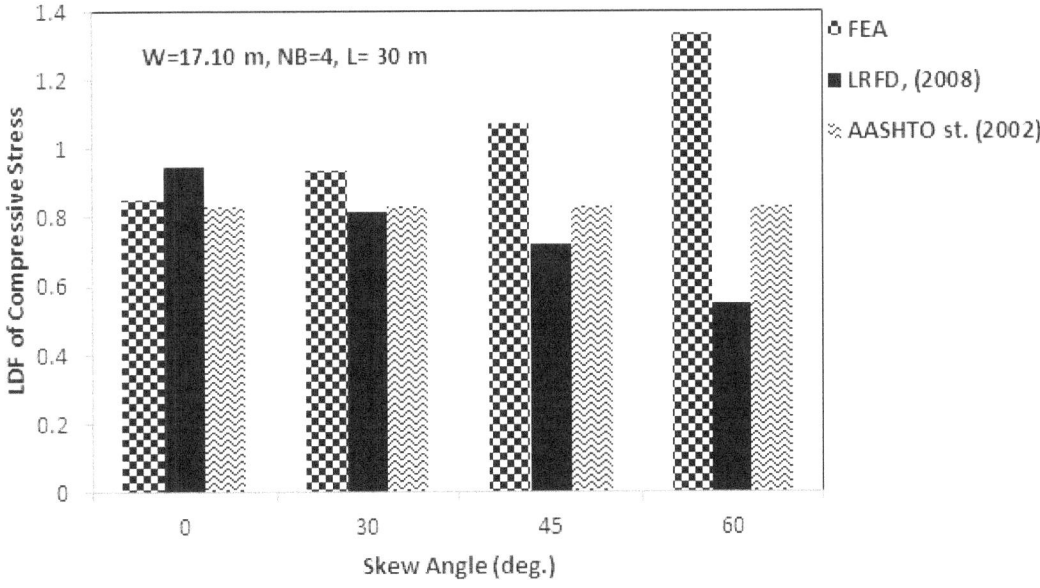

Figure 2.21 Live Load Distribution Factor form various Analytical Method for Bridge with Span Length of 30 m

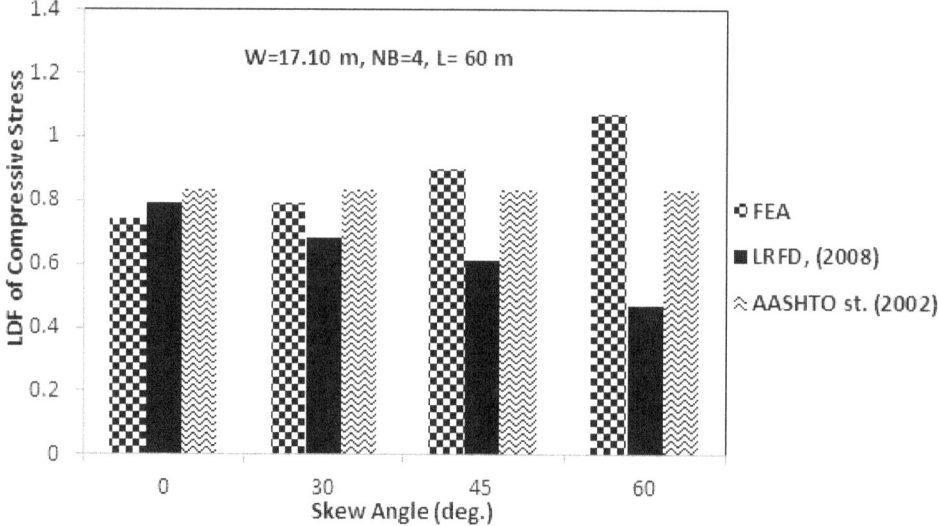

Figure 2.22 Live Load Distribution Factor form Various Analytical Method for Bridge with Span Length of 60 m

It is observed that distributin factor for compressive stress obtained from finite element analysis were significantly higher than those obtained from LRFD formulas, and AASHTO (2002) standard specification. For instance, the difference between the compressive stress distribution factor obtained from FEA and LRFD formulas were up to 30%, for bridge with skew angle of 45° and span lengths of 30 m and 60 m. In addition, the difference between LDF for compressive stress determined from FEA and AASHTO (2002) standard specification were 7% and 25%, repetitively, for four-box bridges with span length of 30 m and 60 m. The variation of live load distribution factor of compressive stress obtained from rigorous method and LRFD formulas has opposite trend. Thus, it was essential to develop new equations for live load distribution factor of compressive stress capable to predict the accurate values.

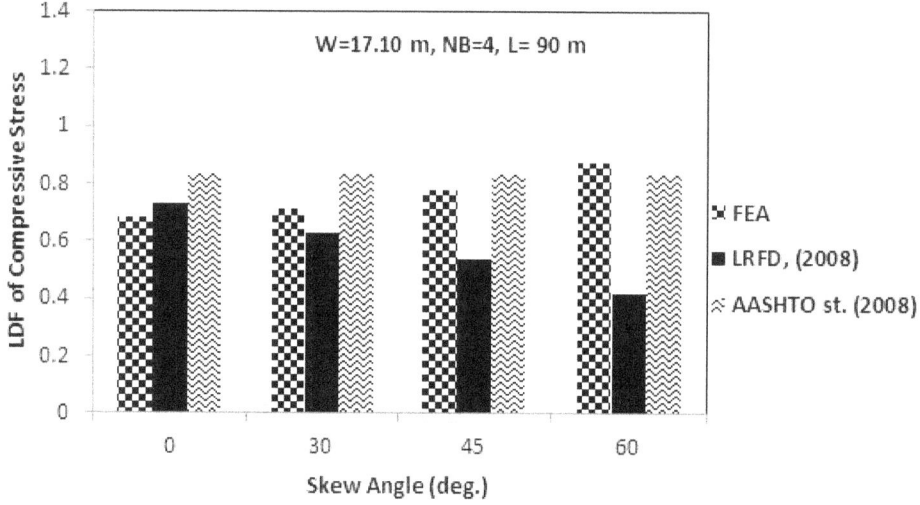

Figure 2.23 Live Load Distribution Factor form Various Analytical Method for Bridge with Span Length of 90 m

2.5   Live Load Distribution Factor of Deflection

It the following sections, the effect of various parameters on distribution of maximum deflection were considred in the same methods as described for stress distribution factor of bridges.

2.5.1   Effect of Span Length

Distribution factor for maximum deflection versus the span length for three, four and six box-girder bridges were plotted in the Figure 2.24 to Figure 2.26.

Clearly, the distribution factor for maximum deflection decreased with increase the span length of bridges. For straight bridges and bridge with skew angle less than 30°, the live load distribution for maximum deflection decreased by almost 18.80% and 12.80% for bridges with skew angle of 60, respectively. Thus, it can be concluded that the span length is a critical parameter affecting the distribution factor of maximum deflection.

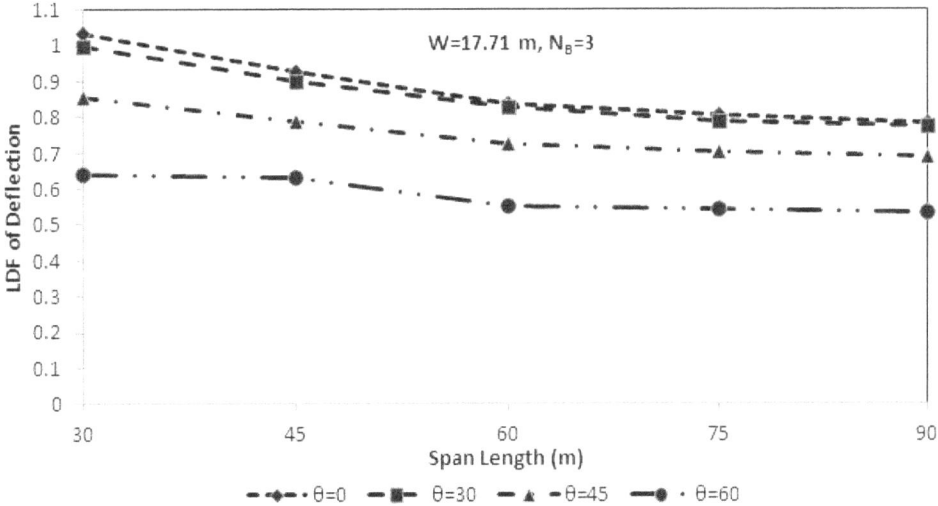

Figure 2.24 Live Load Distribution Factor of Maximum Deflection vs. Span Length for Three Boxes, Four Lane Loads Bridges

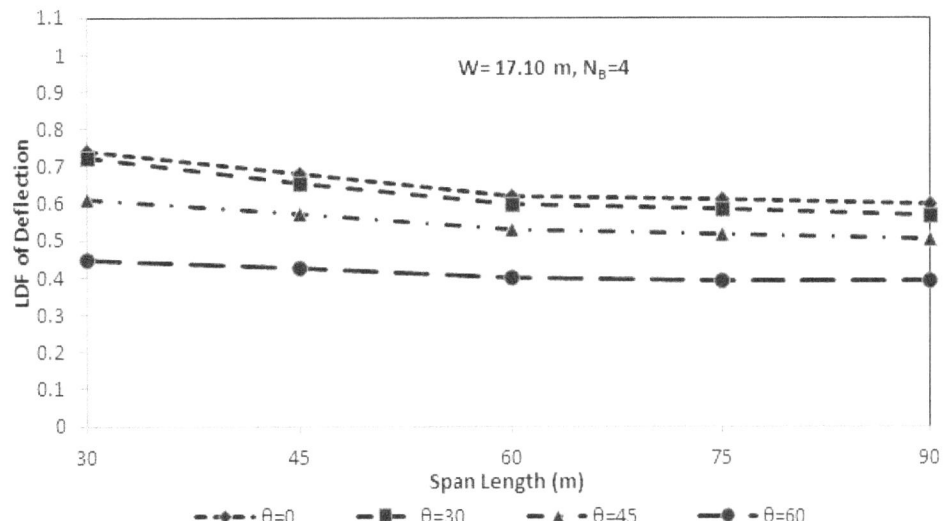

Figure 2.25 Live Load Distribution Factor of Maximum Deflection vs. Span Length for Four Boxes, Four Lane Loads Bridges

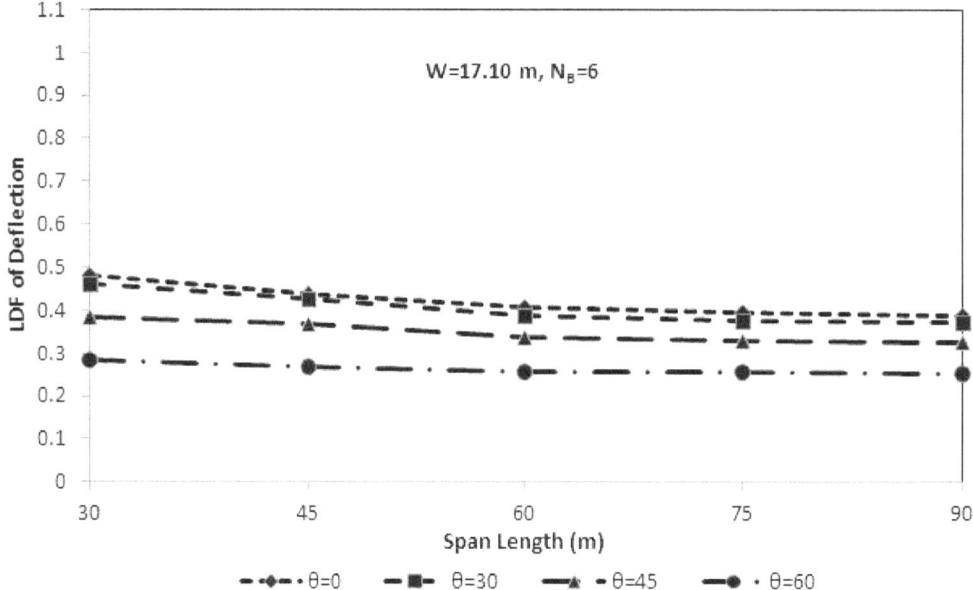

Figure 2.26 Live Load Distribution Factor of Maximum Deflection vs. Span Length for Six Boxes, Four Lane Loads Bridges

2.5.2 Effect of Number of Lane Loads

The relationship between the distribution factors for maximum deflection versus the number of lane loading for three box-girder bridges with span length of 30 m and 60 m and skew angle ranged from 0 to 60° are shown in Figure 2.27 and Figure 2.28.

It can be observed that the distribution factor for maximum deflection increased when the number of lanes increased. The effect of number of lanes in more significant is short span bridges so that for a straight bridges with span length of 30 m, the distribution factor of maximum deflection increased from 0.42 to 0.59 by almost 28%, however, that is approximately 8% for bridge with span length of 60 m. Based on the finding results, it can be concluded that number of lanes is a main parameter for distribution factor of maximum deflection.

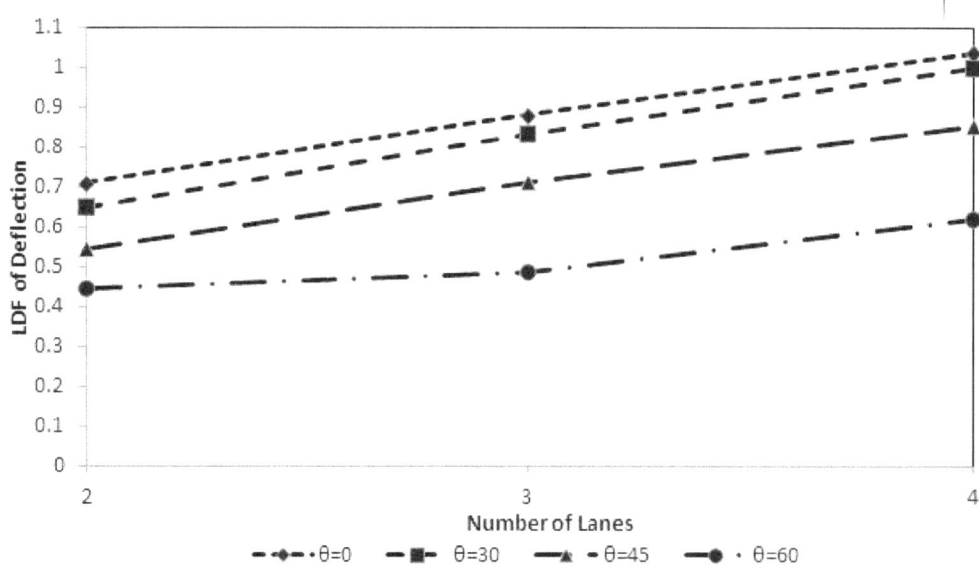

Figure 2.27 Live Load Distribution Factor of Maximum Deflection vs. Number of Lane Loads for 30 m Span Length Bridges

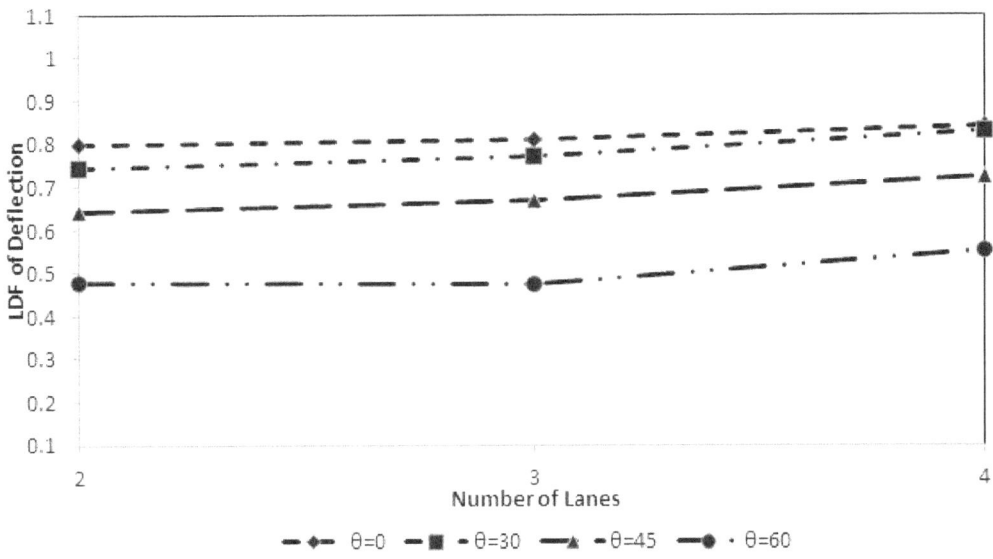

Figure 2.28 Live Load Distribution Factor of maximum deflection vs. Number of Lane Loads for 60 m Span Length Bridges

2.5.3 Effect of Number of Boxes

Figure 2.29 through Figure 2.31 indicate the results of the effect of number of boxes on live load distribution factor for maximum deflection for bridge with span length varied as 30 m, 60 m to 90 m, and skew angle ranged from 0 to 60°.

It was revealed that the distribution factor for maximum deflection decreased with increase the number of boxes by almost 52%. The distribution factor for straight and low skewed bridges is higher than those for high skewed bridges. The variation of distribution factor for deflection and number of boxes followed almost the same trend for various span lengths of bridges.

Thus, the obtained results indicated that the effect of number of boxes should be considered in developing the proposed equations for live load distribution factor of maximum deflection.

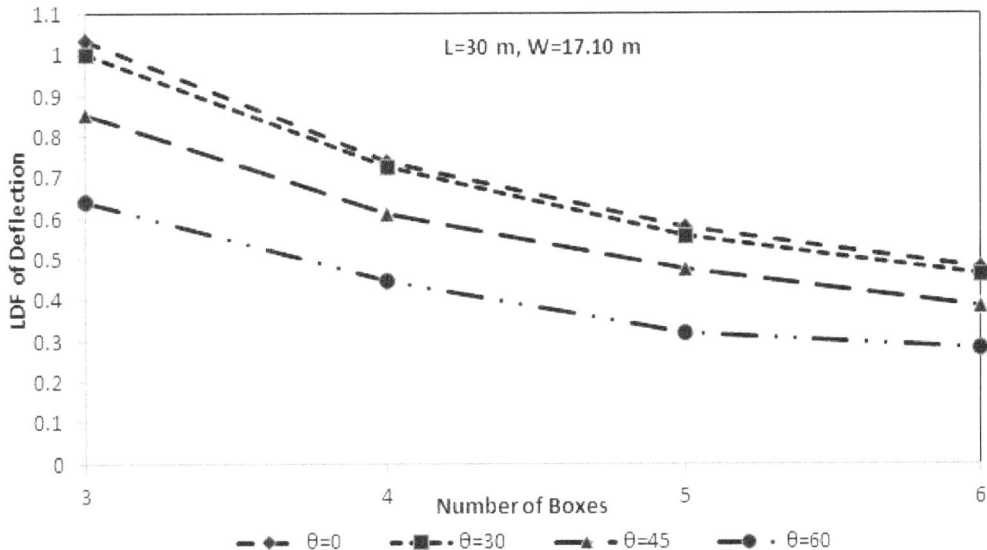

Figure 2.29 Live Load Distribution Factor of Maximum Deflection vs. Number of Boxes for Bridge with 30 m Span

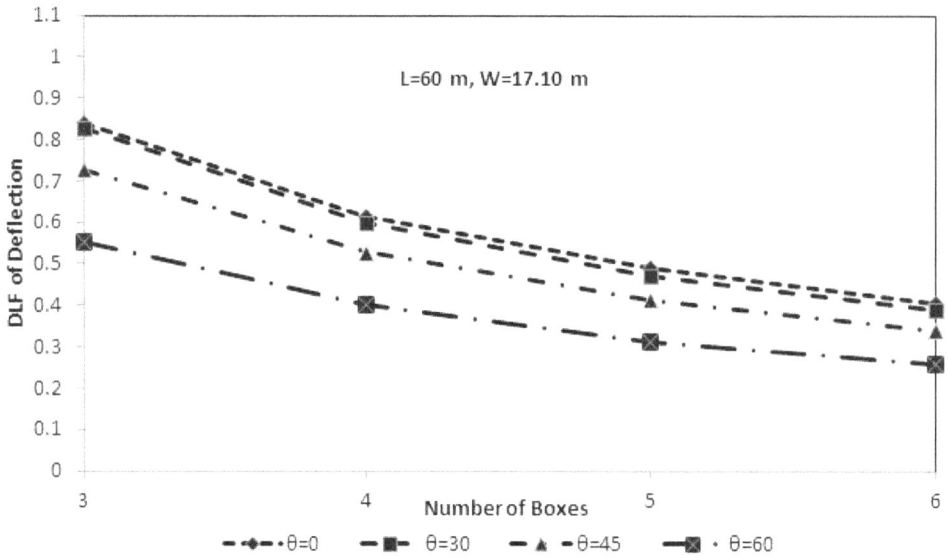

Figure 2.30 Live Load Distribution Factor of Maximum Deflection vs. Number of Boxes for 60 m Bridges

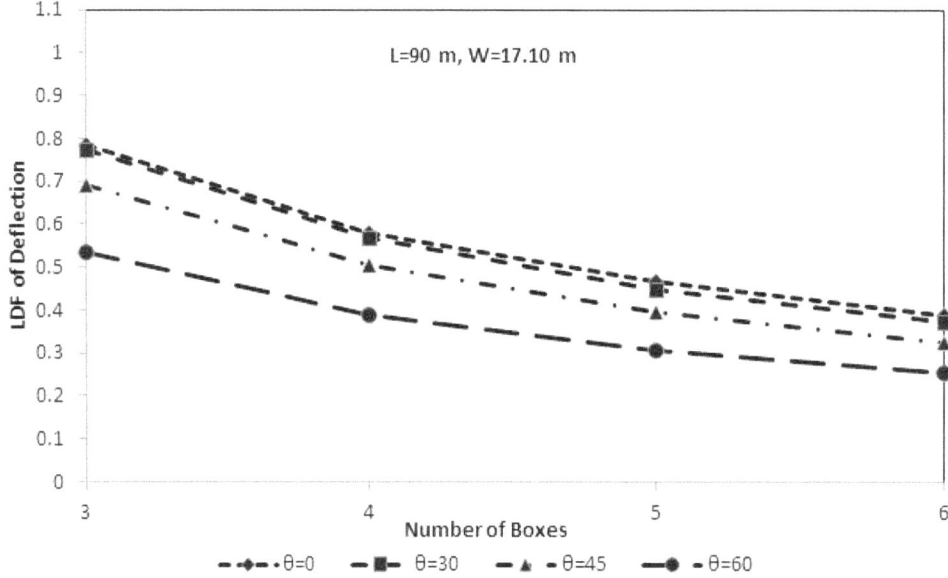

Figure 2.31 Live Load Distribution Factor of Maximum Deflection vs. Number of Boxes for 90 m Span Bridge

2.5.4   Effect of Skew Angle

The relationship between the distribution factor of maximum deflection and the skew angle was plotted in Figure 2.32 and Figure 2.33, for three and six boxes bridges, with span length of 30 m, 60 m and 90 m.

The finding results in graph indicated that the distribution factor of maximum deflection decreased when the skew angle increased. It also revealed that the variation of live load distribution factor for maximum deflection was almost uniform for bridge with skew angle less than 30°.

For bridge with span length of 30 m, the distribution factor changed by almost 40% that indicated the effect of skew angle should be taken into account in extending proposed equation for maximum deflection distribution factor of multicell box-girder bridges.

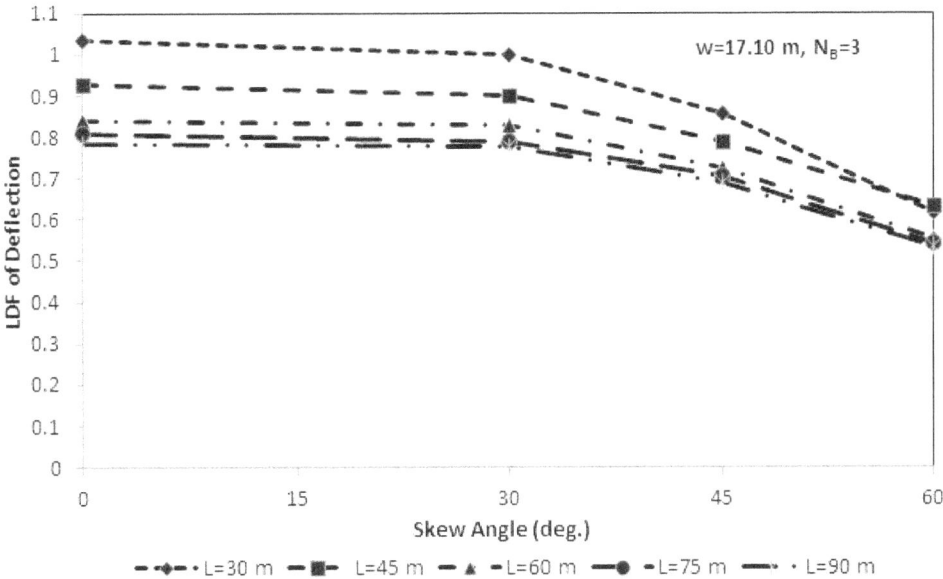

Figure 2.32 Live Load Distribution Factor of maximum deflection vs. Skew Angle for Three-Box bridges

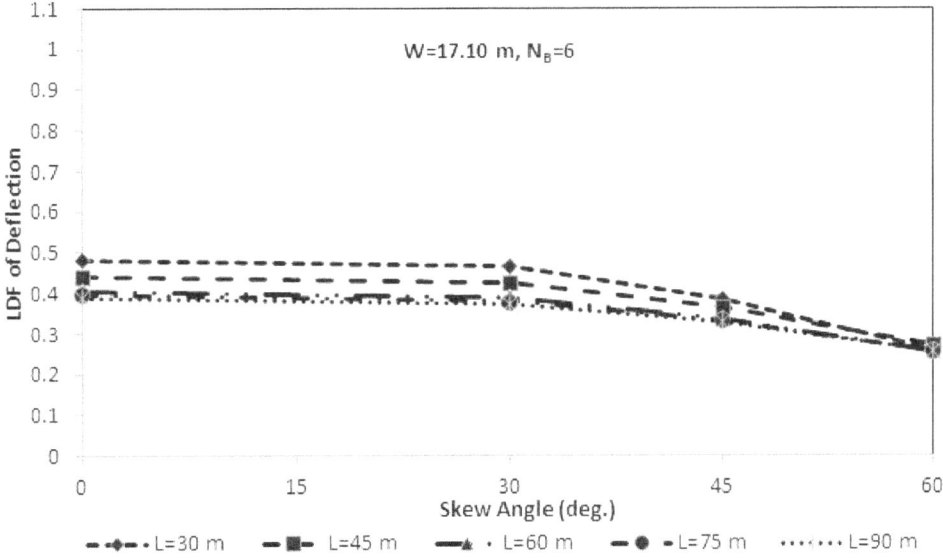

Figure 2.33 Live Load Distribution Factor of maximum deflection vs. Skew Angle for Six-Boxes Bridges

2.6 Empirical Equations for Live Load Distribution Factor

Based on the results obtained from an extensive study on 280 prototype multicell box-girder bridges, empirical equations were developed for the distribution factor of tensile stress, compressive stress as well as maximum deflection. The empirical equations were deduced in term of the main parameters which obtained through the parametric study. The LRFD live load was considered in this study. Using a statistical computer package for best fit based on the method of least square for nonlinear data, the following equation were derived for different distribution factors. It should be noted that the impact factor was not incorporate in these equations, but the multilane presence factor according to AASHTO LRFD (2008) specification were taken into account in the empirical equations.

2.6.1   Development of Equation for Distribution Factor of Compressive Stress

This approach that was already described was used to determine the distribution factor of compressive stress, as following:

Samaan et al. (2002) obtained following empirical formulas for live load distribution factor compressive ($D\sigma_n$) for straight steel spread open-box girder bridges:

$$D\sigma_{ns} = \frac{0.698 \times L^{0.06} \times N_L^{0.70}}{N_B^{0.90}} \text{ (SI Unit)} \qquad (2.1)$$

Where N_B and N_L stands for number of boxes and number of lane loads respectively.

The key parameters of stress distribution factor for superstructures, steel spread open-box girder and multicell box-girder bridge, in term of tensile stress were similar. However, Samaan et al. (2002) did not considered the effect of skewness of distribution factor. Statistical analysis was used to obtain multiplying factor applied to Eq. (2.1), in order to account the distribution factor for compressive stress $D\sigma_n$ of multicell box-girder bridges, as described in the following:

In the beginning, the distribution factor for compressive stress of Eq. (2.1) was simply multiplied by a correction factor (F_N). Accordingly, the negative stress distribution factor was expressed as:

$$D\sigma_{ne} = F_N \times D\sigma_{ns} \qquad (2.2)$$

Where

$$F_N = a \times f(L^{b1},\ N_L^{b2},\ N_B^{b3},\ \cos\theta^{b4}) \qquad (2.3)$$

It is assumed F_N was an exponential function of the form ax^b, where "x" is the value of the given parameter and constants "a" and "b" are to be obtained via regression analysis using the data calculated from finite element analysis.

To determine the constants, the ratio R1 of the negative stress distribution factor calculated from finite element analysis $D\sigma_n$, to that obtained from Eq. (2.1) were plotted as a function of the span length L, as presented in Figure 2.34.

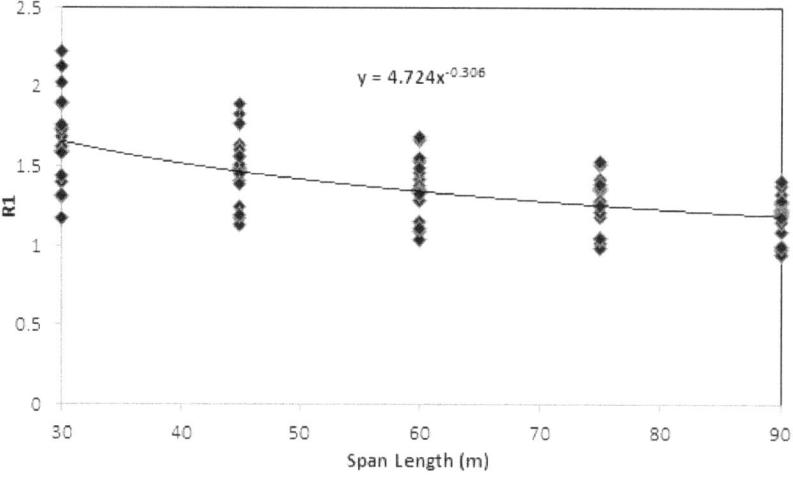

Figure 2.34 Effect of R1 to Span Length

Using the minimum least square fit on the span length as presented in Figure 2.33, the following relationship was obtained:

$$R1 = 4.72 \times L^{-0.306} \qquad (2.4)$$

The Eq. (2.4) determined the ratio of distribution factor obtained using finite element analysis to that calculated from Eq. (2.1). Then term $L^{-0.306}$ in Eq. (2.4) represents the term L^{b1} in Eq. (2.3), so b1=-0.306.

The proof of the scatter present in Figure 2.34 was the absence of other main parameters, number of boxes N_B, and number of lanes N_L. This deviation was removed by considering the effect of other key parameters. For this purpose, R2 was first introduced as a following:

$$R2 = \frac{D\sigma_n}{R1 \times D\sigma_{ns}} \qquad (2.5)$$

where R2 indicated the ratio of the distribution factor for tensile stress obtained from finite element method to those from Samaan formula (Eq. 2.1) modified with respect to L. In Figure 2.35, the ratio R2 was plotted as a function of number of lane loaded N_L.

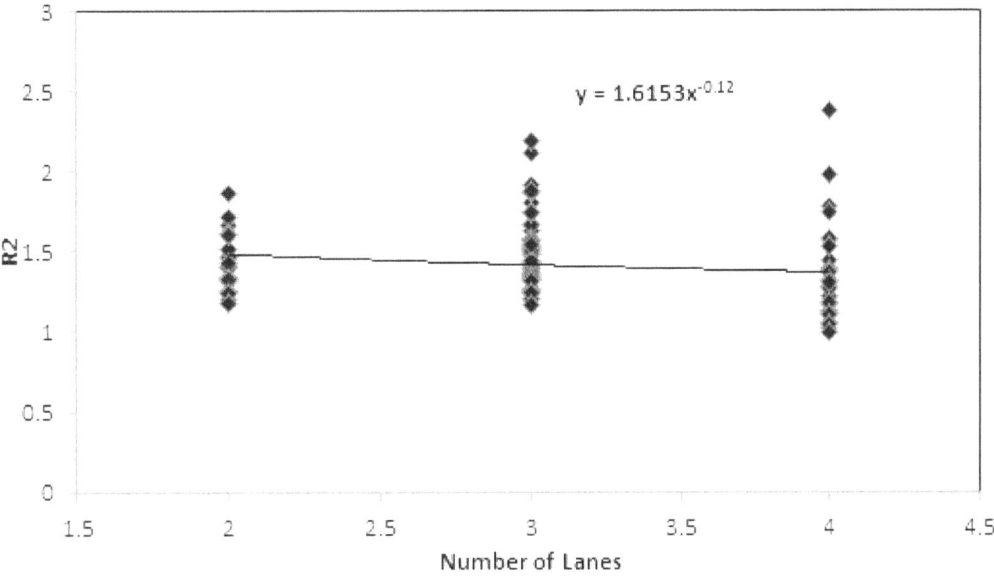

Figure 2.35 Effect of R2 to Number of Lane Loaded

Then the minimum least square fit of the data presented in Figure 2.34 were used to determine the following relationship:

$$R2 = 1.615 \times N_L^{-0.12} \qquad (2.6)$$

The term $N_L^{-0.12}$ in Eq. (2.6) represented the term N_L^{b2} in Eq. (2.3). Thus $b_2 = -0.12$.

The remaining parameter was obtained as a procedure similar to that described above. The relationship between compressive stress and number of boxes was obtained as following;

$$R3 = 1.233 \times N_B^{-0.18} \qquad (2.7)$$

where R3 was determined in the same way as that used to obtain R2. Thus the parameter b2 in Eq. (2.3) was selected equal b3=-0.18. The effect of skewness was described in term of cosine. The relationship between term R4 (which obtained similar to R2) and skew angle was described as following:

$$R4 = 0.638 \times \cos\theta^{-0.40} \qquad (2.8)$$

Thus, term b4 in Eq. (2.3) was obtained equal b4=-0.40. Similarly the constant "a" in Eq. (2.3) was derived by multiplying the b1 to b4 obtained from Eq. (2.4) to Eq. (2.8), such that $b1 \times b2 \times b3 \times b4 = 1.233 \times 4.724 \times 1.650 \times 0.638 = 6.130$. Finally, Eq. (2.7) presented the final form of the correction factor F_N;

$$F_N = \frac{6.130}{N_B^{0.18} \times N_l^{0.12} \times L^{0.3} \times \cos\theta^{0.4}} \qquad (2.7)$$

Thus, the proposed equation for compressive stress distribution factor of multicell box-girder bridges, in the same fashion of proposed equation by Samaan et al (2002), was generated by combining Eq. (2.1), Eq. (2.2) and Eq. (2.7) as following;

$$D\sigma_{ne} = \frac{4.27 \times N_L^{0.58}}{L^{0.24} \times N_B^{1.08} \times \cos\theta^{0.40}} \qquad (2.8)$$

2.7   Live Load Distribution Factor of Tensile Stress

Based on the methods described in Section 3.5 of this research the proposed equations were deduced for distribution factor of tensile stress. The proposed equation was derived by applying the effectiveness of each key parameter on Eq. (2.9), which obtained by Samaan et al. (2002) for straight composite spread box bridges.

$$D\sigma_{po} = \frac{1.255 \times N_l^{0.65}}{L^{0.06} \times N_B} \text{ (SI unit)} \qquad (2.9)$$

The main parameters effecting on tensile stress distribution of live load on bridges are including; span length L, number of lanes N_L, number of boxes N_B, and skew angle of bridges. Using a sensitive statistical analysis the following equation was obtained for live load distribution factor of tensile stress:

$$D\sigma_{po} = \frac{1.737 \times N_l^{0.63} \times \cos\theta^{0.46}}{L^{0.052} \times N_B^{1.181}} \qquad (2.10)$$

A great advantage of Eq. (2.10) in compared to LRFD formulas was that the equation no needs a skew correction factor to consider the effect of skewness.

2.8   Distribution Factor of Maximum Deflection

According to the parametric study, it was concluded that the main parameters for distribution factor of maximum deflection were included; skew angle θ, number of lanes N_L, number of boxes N_B, and span length L.

In the similar way, the minimum least square fit method was applied to deduce a correction factor equation for improving the accuracy of proposed equation (Samaan et al. 2002) for live load distribution factor of maximum deflection. The Samaan equation is as following:

$$D\delta_s = \frac{1.439 \times N_L^{0.65}}{N_B \times L^{0.13}} \quad \text{(SI unit)} \qquad (2.11)$$

The following equation for distribution factor of maximum deflection $D\delta_s$ at the midspan of the skewed MCB bridges:

$$D\delta_s = \frac{3.126 \, N_L^{0.55}}{N_B^{1.35} L^{0.095}} \left(1 - 0.042 \, \tan\theta - 0.108 \, \tan^2\theta\right) \qquad (2.12)$$

The effect of skew angle on maximum deflection distribution factor was expressed as a function of tangent. This equation also can be used to obtain the maximum deflection of straight bridges, it this situation, the second and third phrase into the parenthesis would be zero.

It should be noted that, the proposed Eq. (2.8), Eq. (2.10), and Eq. (2.12) were developed for the case of continuous multicell box-girder bridges with two equal spans. These equations also can be used for simple supported bridges and even multicell box-girder bridges with two unequal continuous spans by taking the longest span length in equations.

2.9   Verification of Proposed Equations

The results of verification analysis of distribution factor for tensile stress, compressive stress and maximum deflection are presented in Figure 2.36 to Figure 2.38.

The thinner line on the graph indicated where the distribution factor obtained from finite element analysis and proposed equations were equal. The points on the graphs present the finite element analysis results versus proposed equation values. The thicker straight line on the plotted graphs is the regression trend line of obtained data which obtain through regression analysis. The aim of regression analysis was to obtain the values of each variable for a trend line that could best fit with the plotted points. If the trend line is close to the unity, it is reveals that the difference between the proposed distribution factor and finite element analysis is small.

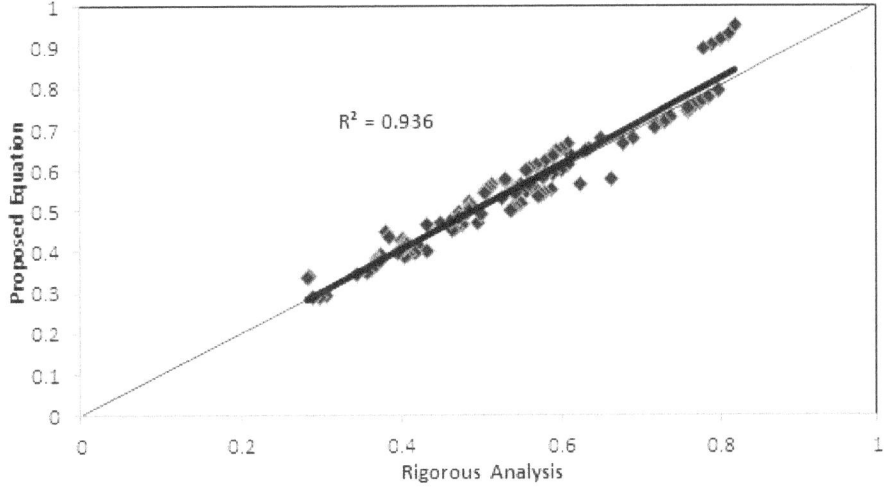

Figure 2.36 Proposed Tensile Stress Distribution Factor vs. Rigorous Distribution Factor of Tensile Stress

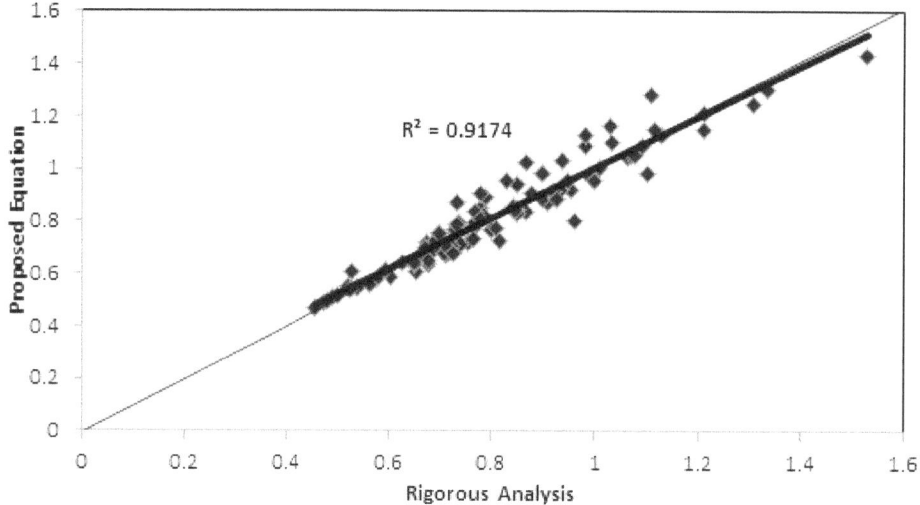

Figure 2.37 Proposed Compressive Stress Distribution Factor vs. Rigorous Distribution Factor of Compressive Stress

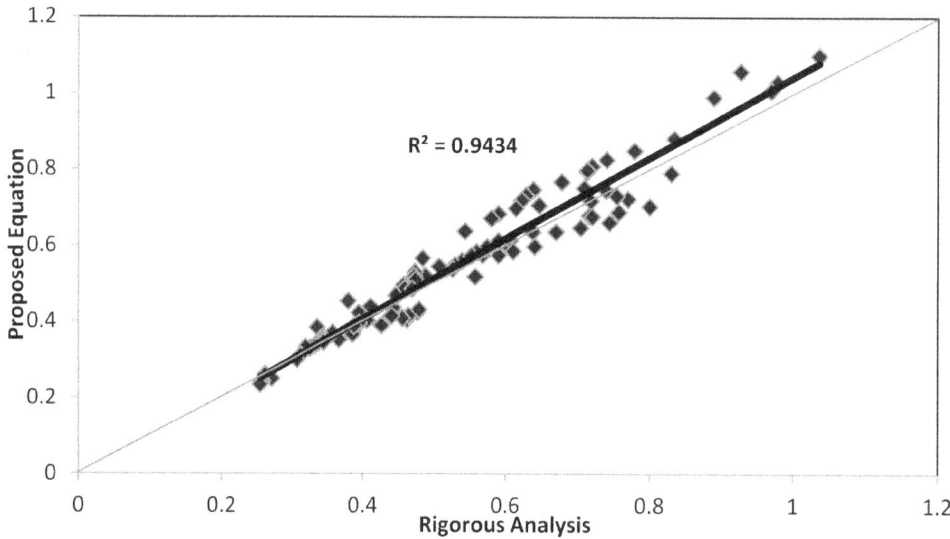

Figure 2.38 Proposed Deflection Distribution Factor vs. Rigorous Distribution Factor of Maximum Deflection

The coefficient of determination, R^2, was calculated using statistical analysis. The coefficient of determination indicates the rate of variance of one parameter that is foreseeable from other parameter. It is a scale that can be employed to calculate how certain one can be estimate from a plotted graph. This factor indicates the percent of values that are the nearest to the best fit line. Therefore, high coefficient of determination, R^2, presents weak variety.

It was observed from Figure 2.36 to Figure 2.38 that the R^2 for tensile stress is almost 0.936, which definite very slight variety of the obtained results. The R^2 for compressive stress and maximum deflection distribution factor were also larger than 0.90 which were excellent.

The average AVG., standard deviation SD., and coefficient of variation COV, were determined for each set of data. The average should be slightly higher than unity to ensure quite acceptable results. The standards deviation must be low that the equations that obtained these ratios are evaluated to be the sufficient values.

Compressive study of live load distribution factors for stress and maximum deflection for straight and skewed multicell box-girder bridges were indicated in Table 2.5 The average values (Proposed equations/Finite element analysis) for tensile stress, compressive stress and maximum deflection of prototype bridges were very close or slightly higher than the unity, which defined that the average of derived equations were approximately equal to the mean of finite element results.

Table 2.5 Comparative Statstics of Proposed Equations

Live Load Distribution Factor (LDF)	Average (AVG.)	Standard Deviation (SD.)	Coefficient of Variation (COV.)
Positive stress	1.023	0.064	0.0623
Negative stress	1.017	0.069	0.0674
Deflection	1.024	0.077	0.0773

Variance is the average of the squared differences between points of values and the mean. The coefficient of variances of 0.062, 0.0674, and 0.0773 determined for live load distribution factor of tensile stress, compressive stress and maximum deflection, which were so excellent (Zhang 2008; Zheng 2009). Standard deviation (SD) is the most common measure of statistical dispersion which indicates how widely the values are spread in a data set, being the square root of the variance. From Table 6.1, the standard deviations were 0.064, 0.069, and 0.077, which indicated that the distribution factors were sufficiently close to mean values.

2.10 Effect of Number of Spans on Live Load Distribution Factor

Figure 2.39 shows the results on the effect of number of spans on lateral load distribution factor of tensile stress, compressive stress and maximum deflection for four boxes bridge when the number of spans was changed from two to four.

It can be seen that with the increase of the number of span length from two to four, the live load distribution factor of tensile stress kept almost constant. The compressive stress distribution factor and distribution factor for maximum deflection were varied by almost 4.2%, when number of spans increased from two to four. Thus, it was concluded that the proposed equations of live load distribution factor of stress and maximum deflection could be used in design of multispan bridges

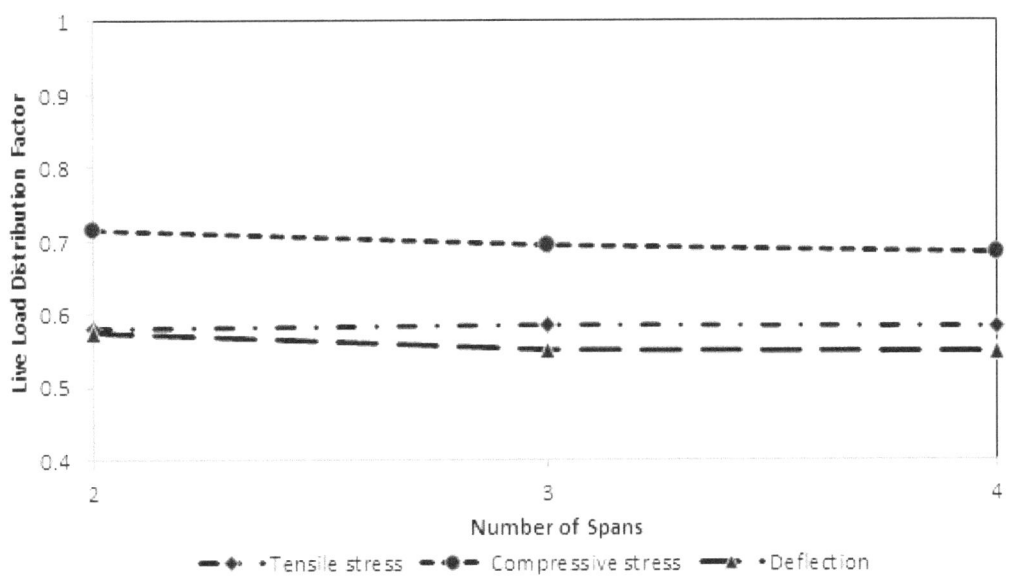

Figure 2.39 Effect of Number of Spans on Live Load Distribution Factor of Four Boxes Bridge with Span Length of 45 m

2.11   Effect of Intermediate Diaphragms on Live Load Distribution Factor

Intermediate diaphragms are essential to maintain the concrete box-shape and to decrease the lateral bending and longitudinal warping stress in box-girder bridges, However, the use of intermediate in box-girder bridges is still a controversial issue. Even if the intermediate diaphragms are needed, current bridge specifications that command provision of intermediate diaphragms do not take into account the influence of diaphragms in lateral load distribution factor of bridges (Chandolu 2005; AASHTO LRFD 2008; Zheng 2009). Evaluating the contribution of diaphragms in lateral load distribution could modify the bridge design process for multicell box-girder bridges. Thus, in this section the effect of intermediate diaphragms on distribution of live loads were considered.

2.11.1   Effect of Intermediate Diaphragms on Stress Distribution Factor

To evaluate the influence of intermediate diaphragms on the bridge actions, a two span four box-girder bridge of 60 m span length, and with skew angle ranged from zero to 60°, was analyzed subjected to LRFD live loading.

The relationship between distribution factors for tensile stress, compressive stress, is indicated in Figure 2.40. It was concluded that setting up the intermediate diaphragm has an insignificant effect on lateral load distribution factor for tensile stress and compressive stress. Therefore, the effect of intermediate diaphragms in developing equation for live load distribution factor of stress can be neglected.

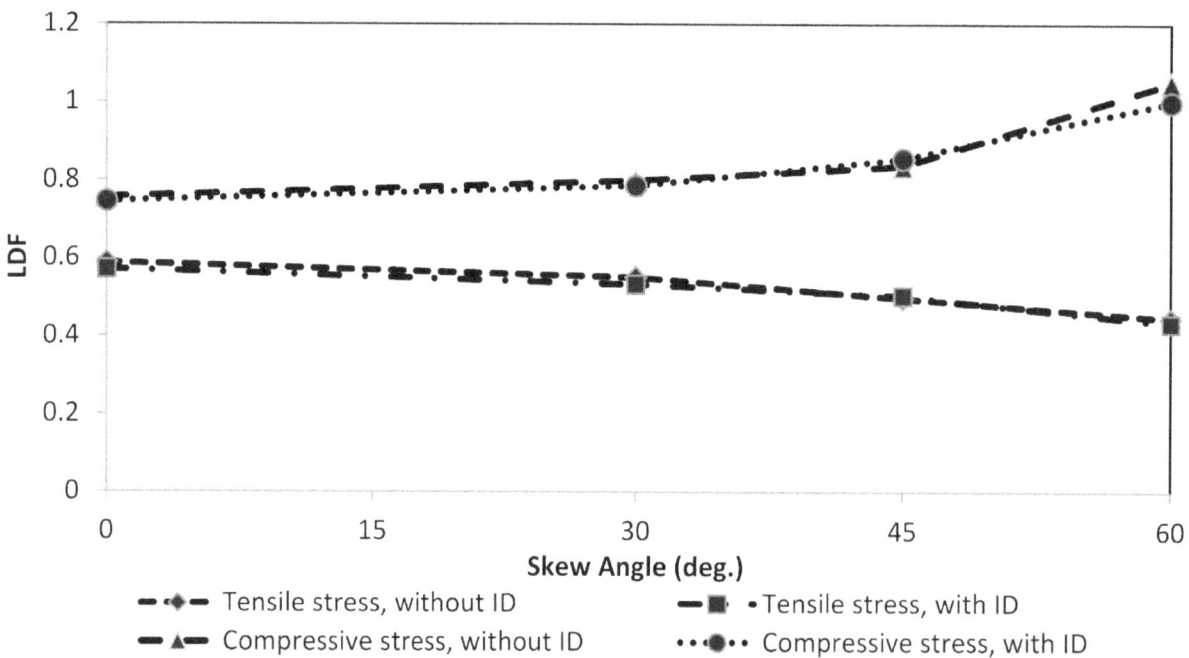

**Figure 2.40 Effects of Intermediate Diaphragms (IDs) on
Lateral Load Distribution of Tensile and Compressive Stress**

In addition, Khaloo & Mirzabozorgi (2003) indicated that the intermediate diaphragms significantly affect live load distribution factor of bending moment; however, it did not take into account in developing the AASHTO LRFD formulas for live load distribution factor. In the following section, the effect of intermediate diaphragms on live load distribution of bending moment was considered.

2.12 Evaluation the Effect of Bridge Parameters on Effectiveness of Intermediate Diaphragms

As mentioned before the intermediate diaphragm has a significant effect on lateral distribution factor of bending moment. On the other hand the effect of intermediate diaphragms did not considered in the development of LRFD formulas for distribution factors. Thus, the main objective of this unit was to develop correction factors for distribution factor to account for the influence of diaphragms in lateral load distribution factor of bending moment.

Based on different recommendations of various regulations on how to use the intermediate diaphragms on bridges, several arrangements diaphragms are considered in this study. In the first arrangement, the bridges are without any intermediate diaphragm (S-1). In the second arrangement, intermediate diaphragms are parallel to the supporting line (S-2). For this system, according to Louisiana bridge design manual (LADOTD 2002), intermediate diaphragms are applied at one-third and two-third of the span lengths.

In the third arrangement, intermediate diaphragms are perpendicular to the longitudinal girders. For this arrangement, three various cases are considered. In the first case, According to (AASHTO 2002), intermediate diaphragm is located at the midspan of bridges (S-3-1). In the second case, the location of intermediate diaphragms are according to Louisiana bridge design manual, (LADOTD 2002) with two intermediate diaphragms located at the one-third and two-third of span lengths (S-3-2). In the third case, intermediate diaphragms are located at spacing 7.60 m (S-3-3). These various systems of intermediate diaphragms arrangements are listed in Table 2.6 It this section the bridge with single span and continuous span were investigated. The properties of single bridges were similar to the described in Chapter V of parametric study.

Table 2.6 Various Arrangements of Intermediate Diaphragm (IDs)

No. of Set	S-1	S-2	S-3-X
Arrangement	No ID	Parallel	perpendicular
Location of ID	-	(1/3 and 2/3) L	X=1 1/2 L X=2 1/3 and 2/3 L X=3 at spacing 25 feet

2.13 Effect of Span Length

The effect of span length on effectiveness of intermediate diaphragms is plotted in Figure 2.41, for distribution factor of bending moment of external and internal girder. It was observed that the presence of intermediate diaphragms significantly affect the variation of live load distribution factor of bending moment so that distribution factor of bending moment for external girder increased with increase the span length.

The graph also indicated that providing the intermediate diaphragms parallel to skew supports, has insignificant effect on distribution factor of both external and internal girders. The slight increase in distribution factor of internal girder evidenced that the effect of intermediate diaphragms on span length of internal girder could be neglected.

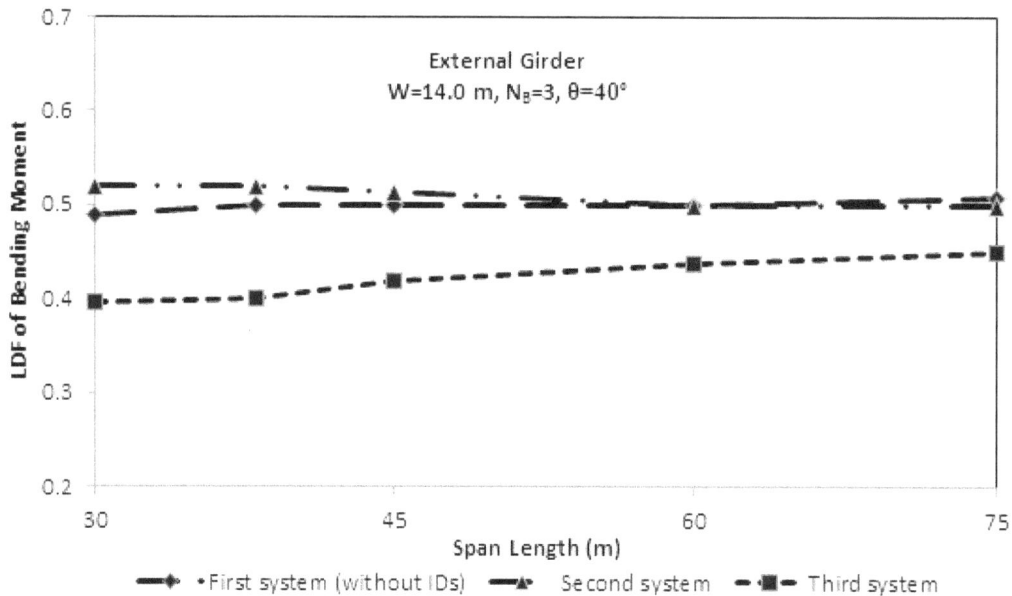

a) External Girder

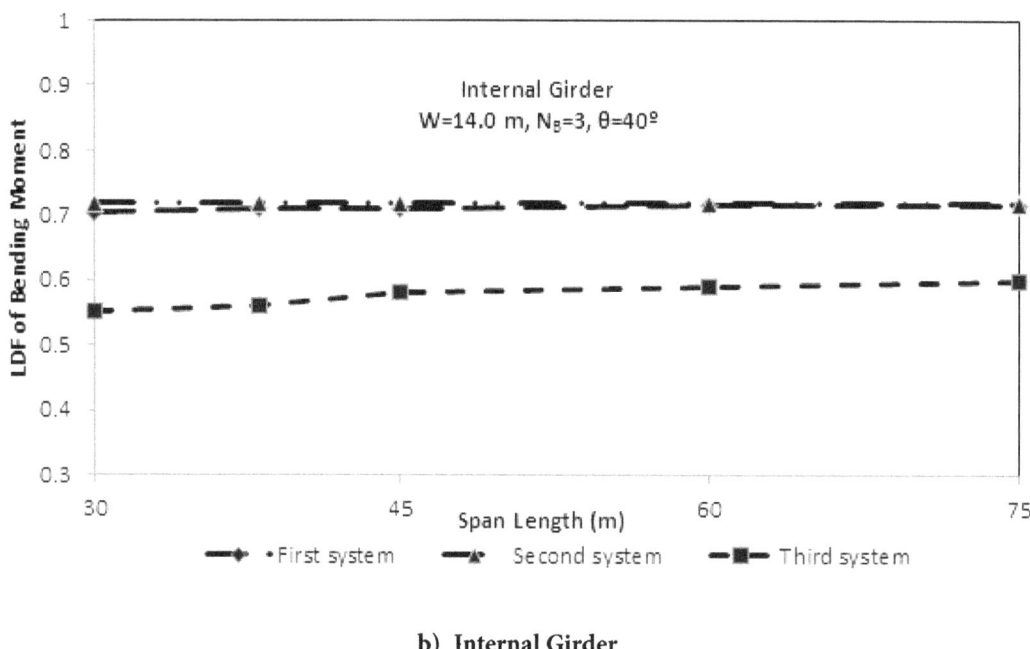

b) Internal Girder

Figure 2.41 Effect of Span Length on Bending Moment Distribution Factor

2.13.1 *Effect of Skew Angle*

The relationship between the skew angle and live load distribution factor for bending moment of external and internal girder are indicated in Figure 2.42, for a three lane bridge with 30 m of span and three box-girders. Three cases of intermediate diaphragms arrangements were evaluated including S-1, S-3-2, and S-3-3. The skew angle ranged from 0 to 50°. The bridge with skew angle of 60° was not considered for bridge with 30 m span length due to limitations of diaphragms modeling with SAP2000 program.

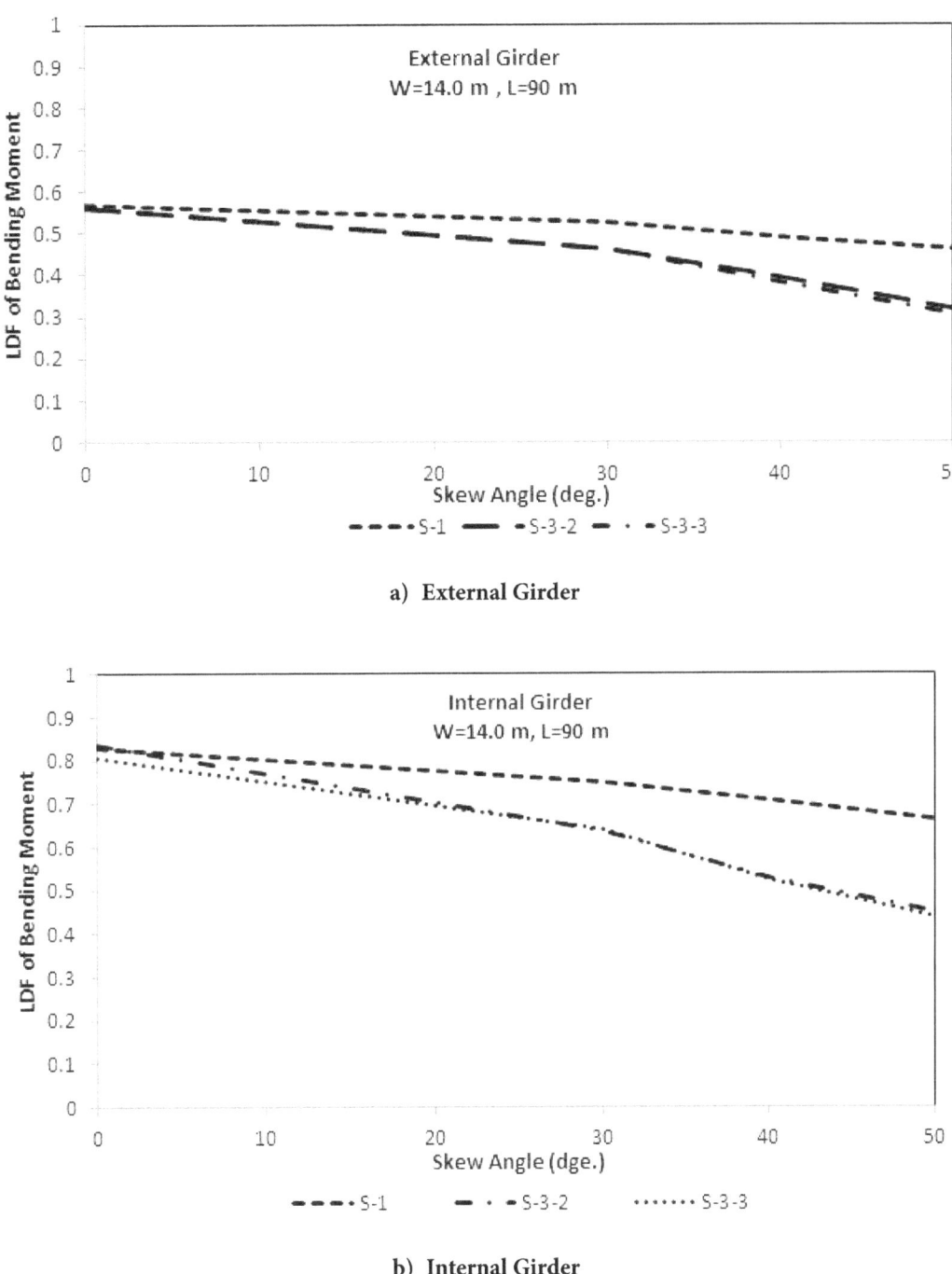

a) External Girder

b) Internal Girder

Figure 2.42 Effect of Skew Angle on Bending Moment Distribution Factor

As expected, the skew angle of the superstructure was the most effective factor on distribution of bending moment of bridges including the intermediate diaphragms. The Figure 2.42 indicates that the live load distribution factor of bending moment decreased when skew angle increased by almost 45.5% and 50% for bridge with two intermediate diaphragms (S-3-2) and that with intermediate diaphragms located at spacing 7.6 m (S-3-3). The reduction was more significant when skew angle exceed from 30°.

The decrease in distribution factors reduced significantly when skew angle exceeded from 30°. It was also revealed that the number of intermediate diaphragms has insignificant effect on distribution of bending moments. Thus, the effect of skew angle should be taken into account in developing the modification factor for effectiveness of intermediate diaphragms on bending moment distribution factor.

2.13.2 Effect of Intermediate Diaphragm Arrangements

The effect of intermediate diaphragm arrangements on live load distribution factor of bending moment for three and four box-girder bridges with skew angle of 45° are shown in Figure 2.43. It is indicated that providing intermediate diaphragms paralleled to support direction (S-2) have an insignificant effect on distribution of bending moment for both external and internal girders, by almost 6.5% and 1.0% for external and internal girder of bridge with four boxes, respectively. In the third system (S-3), intermediate diaphragms greatly influence distribution factors of bending moment by up to 19% in both external and internal girders.

It also revealed from the graphs that the number of intermediate diaphragms in the third models (S-3-1 to S-3-3) had a slight influence on distribution of bending moment. Since presence of intermediate diaphragms increases the cost and time of construction, the cases with less number of intermediate diaphragms are more acceptable.

On the other hand, Yang et al. (2010) concluded that the response of bridge with two intermediate diaphragms (S-3-2) to over-height truck impacts distributes more uniform in the girders, and the area of damage is greatly smaller than the bridge with only one intermediate diaphragms (refer to S-3-1). Hence, the second arrangement (S-3-2) was recommended to secure both economic and safety targets.

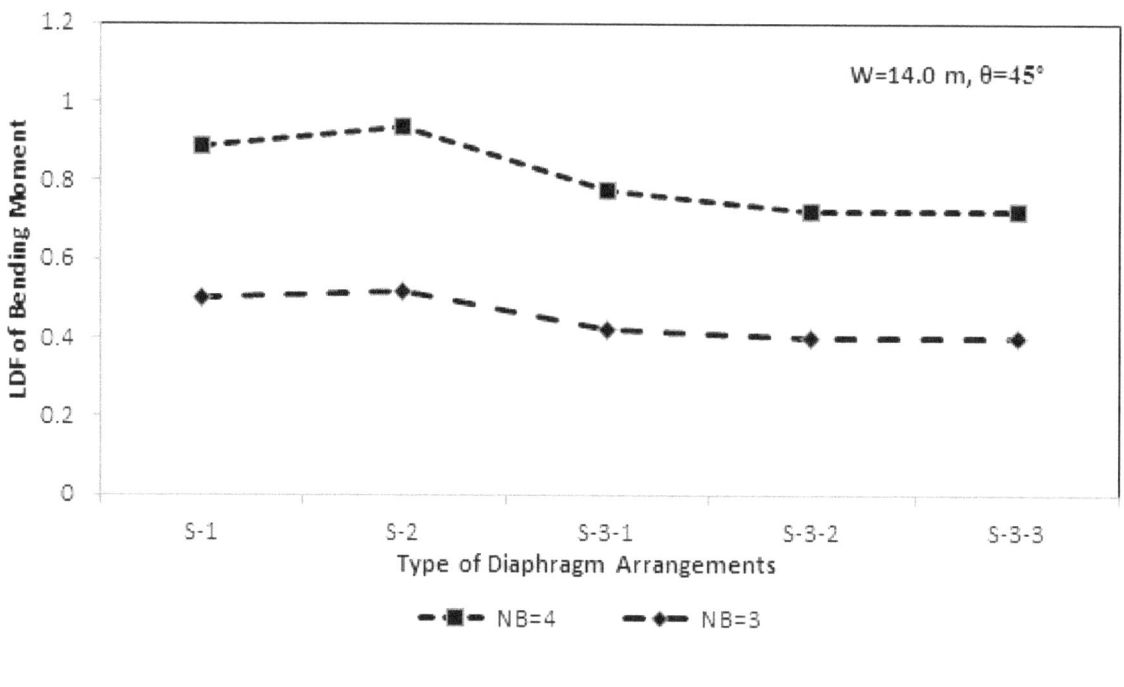

a) External Girder

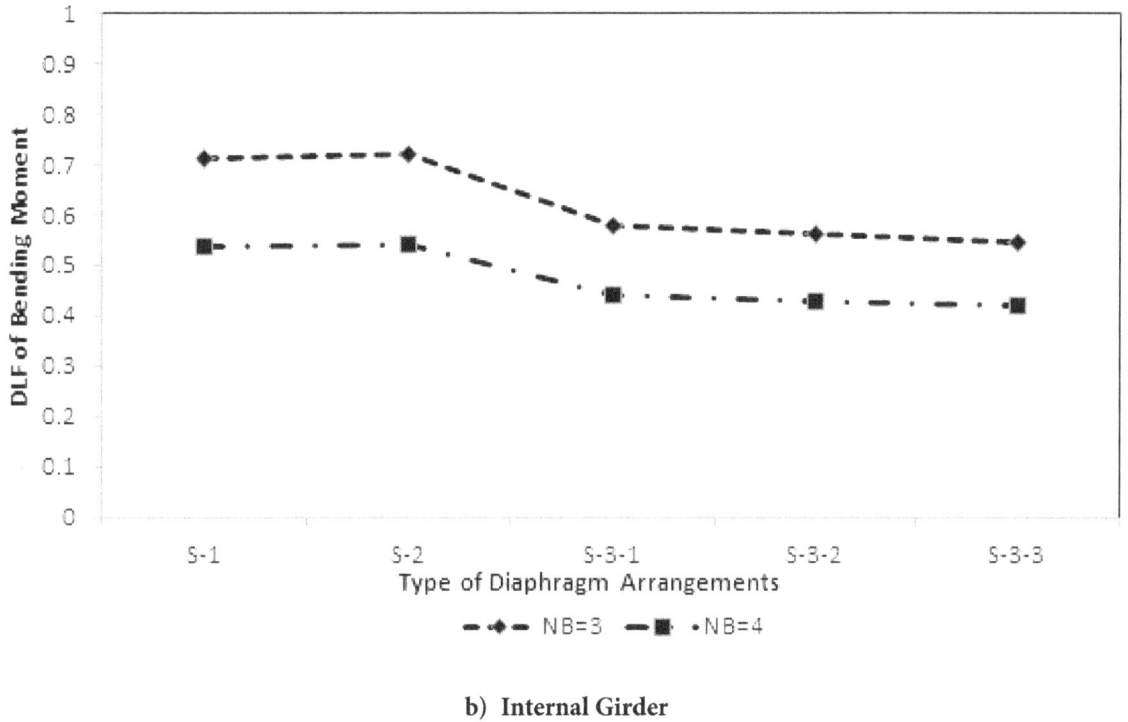

b) Internal Girder

Figure 2.43 Effect of Intermediate Diaphragm Arrangements on Bending Moment Distribution Factor

It should be mentioned that in straight bridges (θ=0°), the influence of intermediate diaphragms on live load distribution factor of bending moment are insignificant, however due to possibility of over-height truck impacts and to provide lateral supports to deck components under construction loads, intermediate diaphragms should be used.

2.13.3 Formulation of Effectiveness of Intermediate Diaphragms

The AASHTO LRFD specifications (2008) developed new formulas for live load moment without considering the effect of intermediate diaphragms. Form previous section, it was observed that the intermediate diaphragms widely influence on live load distribution factor of multicell box-girder bridges. In this unit using statistical analysis, several proposed equations were deduced to take into account the effectiveness of intermediate diaphragms in moment distribution factor of multicell box-girder bridges.

First, the percentage change in live load distribution factor of bending moment R_d was gauged by the following relationship to evaluate the influence of intermediate diaphragms;

$$R_d = \frac{DF_{ND} - DF_{WD}}{DF_{ND}} \times 100 \qquad (2.13)$$

In above equation, DF_{WD} and DF_{ND} are distribution factor for bridges with and without intermediate diaphragms, respectively. Based on parametric study, only span length and skew angle of bridges were considered as possible variable affecting intermediate diaphragms effectiveness.

2.13.4 Development of R_d Equation for Internal Girders

To develop the R_d equation for internal girders, a relationship was obtained between span length and diaphragms effectiveness as shown in Figure 2.44. It was observed that at each skew angle, R_{d1} values of each skew angle lie fairly on a straight line and the all graph lines are quite parallel to each other. Using the minimum least square fit analysis for bridges with skew angle equal 40°, the following relationship was obtained;

$$R_{d1} = 26.14 - 0.148 \times L \qquad (2.14)$$

To determine the effect of skew angle on R_d and to reduce the amount of scatter values between graphs in Figure 2.44, the ratio R_1 was obtained from Eq. (2.15):

$$R_1 = \frac{R_d}{26.14 - 0.148L} \qquad (2.15)$$

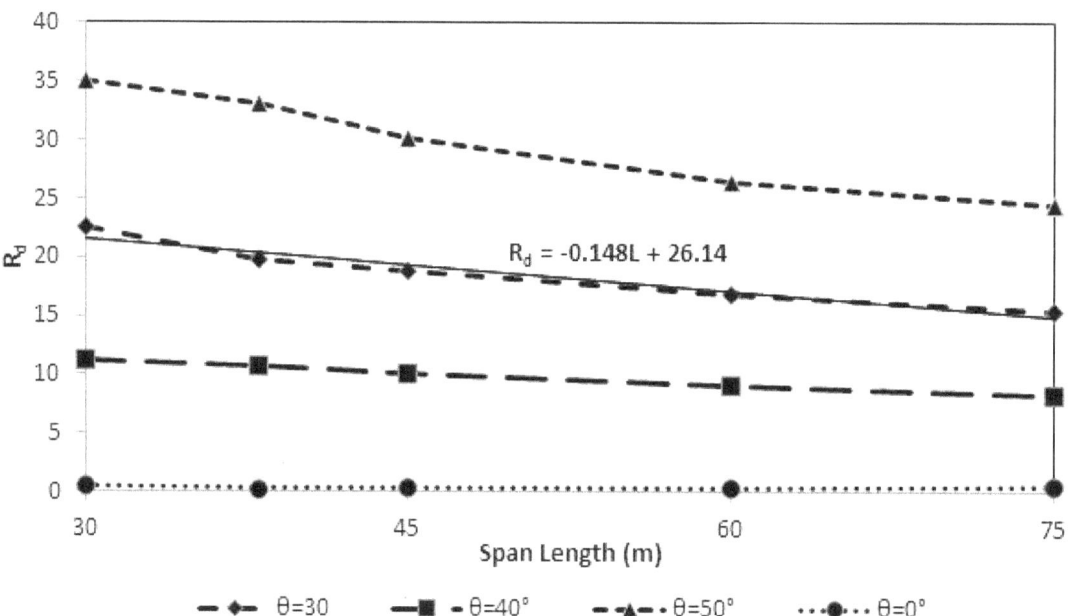

Figure 2.44 **Effect of Span Length on Intermediate Diaphragms Effectiveness of Internal Girder**

A plot was drawn between skew angle and R_1 in Figure 2.45. Then a relation was deduced using of minimum least square fit analysis as follows:

$$R_1 = 1.318 \times \tan\theta - 0.07 \qquad (2.16)$$

Finally, the following equation was deduced for determining R_d value for internal girder:

$$R_{dn} = (26.14 - 0.148L) \times (1.318 \times \tan\theta - 0.07) \qquad (2.17)$$

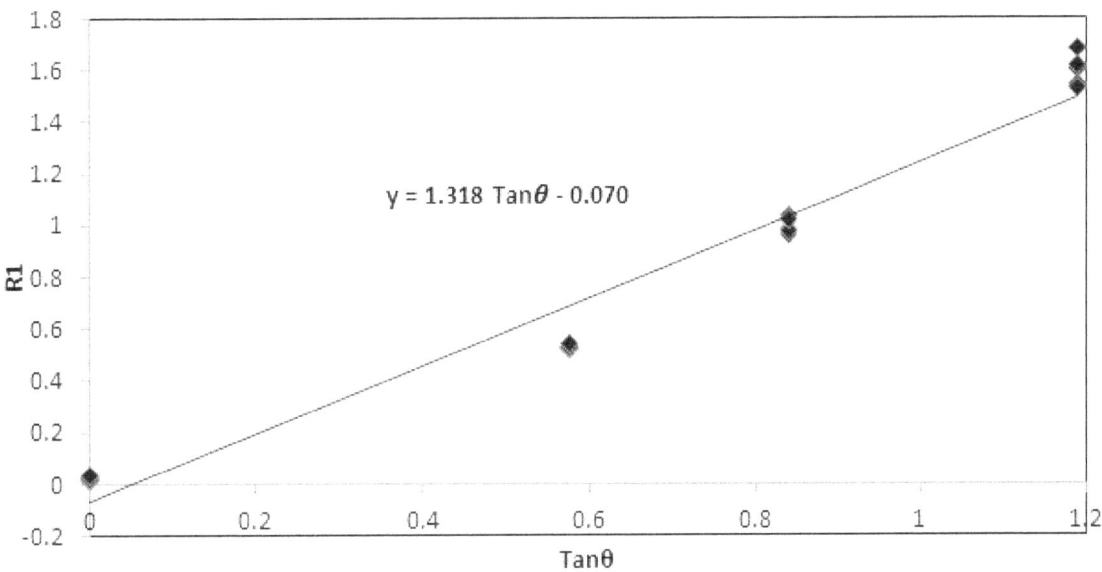

Figure 2.45 Effect of Skew Angle on R1

2.13.5 R_d Equation for External Girders

To find an equation for R_d of external girders, it was assumed that R_d equation could be obtained in the similar form to Eq. (2.16). Accordingly, using a process similar to what was explained in the previous section, the following equation was deduced:

$$R_{dx} = (25.20\text{-}0.21{\times}L){\times}(0.8{\times}\tan\theta\text{-}0.028) \qquad (2.18)$$

Finally, bending moment distribution factor equation of multicell box girder bridges with taking into account diaphragm effectiveness was given by the following expression:

$$DF_{WD} = \left(1\text{-}\frac{R_d}{100}\right){\times}DF_{ND} \qquad (2.19)$$

where DF_{WD} is the distribution factor of bending moment of bridge with intermediate diaphragms, and DF_{ND} was bending moment distribution factor of bridge without intermediate diaphragms. If DF_{ND} in Eq. (2.18) was calculated from finite element analysis, the corresponding DF_{WD} and R_d were named as "FEA Based Equation", and if that was obtained from formulas specified in the current AASHTO LRFD specifications, the corresponding DF_{WD} and R_d were named as "LRFD Based Equation".

2.13.6 R_d Equations for AASHTO LRFD Specification

Previous study by Song et al. (2003) indicated that the current AASHTO LRFD formulas predict too conservative results for moment distribution factors due to neglecting the effectiveness of diaphragms on distribution of bending moment.

To improve the accuracy of LRFD formulas for distribution of bending moment, a statistical analysis based on best minimum least square fit of data as described in Chapter III was employed on three and four box-girder bridges, and the following equations were deduced for internal and external girder, respectively:

$$DF_{in} = \left(0.243 \times L^{0.25} - \frac{A \times B}{100} \right) \times LLDF_{in} \qquad \text{For } \theta \geq 20 \qquad (2.20)$$

$$DF_{ex} = \left(0.75 - \frac{C \times D}{100} \right) \times (1.30 - 0.033 \times W_e) \times LLDF_{ex} \qquad (2.21)$$

In proposed equations, LLDF is the live load distribution factor of bending moment, and the subscript 'ex' and 'in' stand for external and internal girders, respectively. W_e is the half of bridge and total width of overhang. The parameters A, B, C, D are described as following:

$$A = 26.14 \times L - 0.148 \times L^{1.25} \qquad (2.22)$$

$$B = 0.320 \times tan\theta - 0.017 \qquad (2.23)$$

$$C = 18.90 - 1.55 \times L \qquad (2.24)$$

$$D = 0.80 \times tan\theta - 0.028 \qquad (2.25)$$

2.14 Verification of Empirical Equation

The results of validation analysis of R_d equations which developed to consider the effectiveness of intermediate diaphragms on live load distribution factor of bending moment of external and internal girder of multicell box-girder bridges were shown in Figure 2.46 to Figure 2.49. The figures indicate the relationship between finite element results and those from equation with and without considering the effect of diaphragms.

It can be observed that the coefficient of determination, R^2, for all proposed equations for live load distribution factor of bending moment based on FEA and based on LRFD, respectively, were higher or slightly lower than 0.90 which was excellent and indicated a low variety of the finite element analysis results. It was also concluded that the effectiveness of intermediate diaphragms is more significant on internal girders. It was due to distribution of higher bending moment subjected to vehicle weight on internal girders. The external girder was more subjected to torsional moment due to eccentricity of truck loading and external girder.

By comparing the R^2 of bridge for before and after considering the effectiveness of intermediate diaphragms, it was revealed that the proposed equations estimate more efficiently account the effect of diaphragms on live load distribution factor of bending moment.

The standard deviation (SD.), average (AVG.) and variation were presented in Table 2.3 for each set of ratios between equations and finite element analysis results (Equation/FEM).

It can be observed that the average values for both external and internal girder were very close to unity which was so excellent. In Table 2.7, the standard deviations of 0.035 to 0.100 revealed that distribution factor of bending moment were sufficiently close to mean values (Zhang 2008). It should be noted that the proposed equations for effectiveness of intermediate diaphragms were deduced with analysis only three and four box-girder bridges due to insignificant effect of number of boxes on effect of diaphragms.

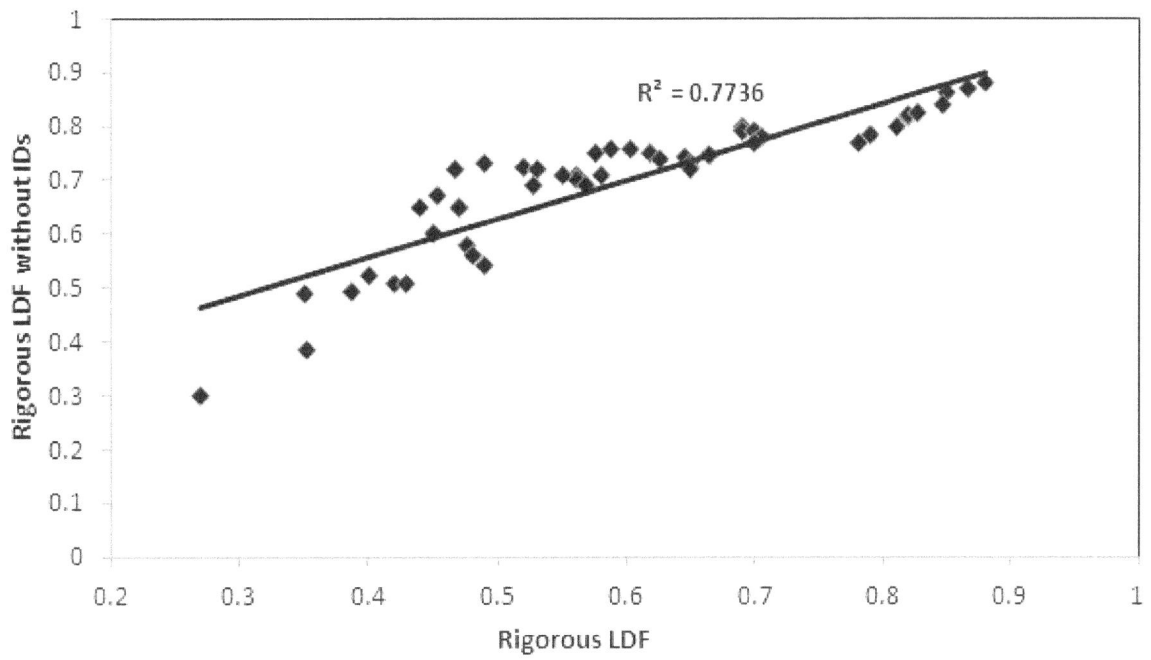

a) Without considering IDs effectiveness

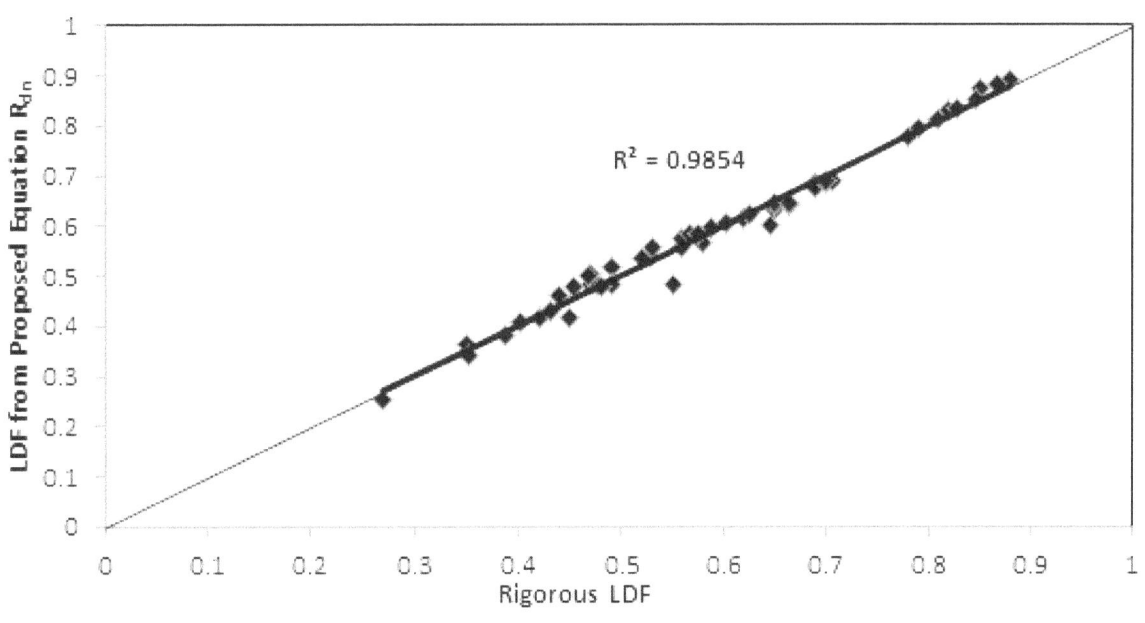

b) With considering IDs effectiveness

Figure 2.46 Live Load Distribution Factor from Rigorous Analysis and Proposed Equation for Internal Girder

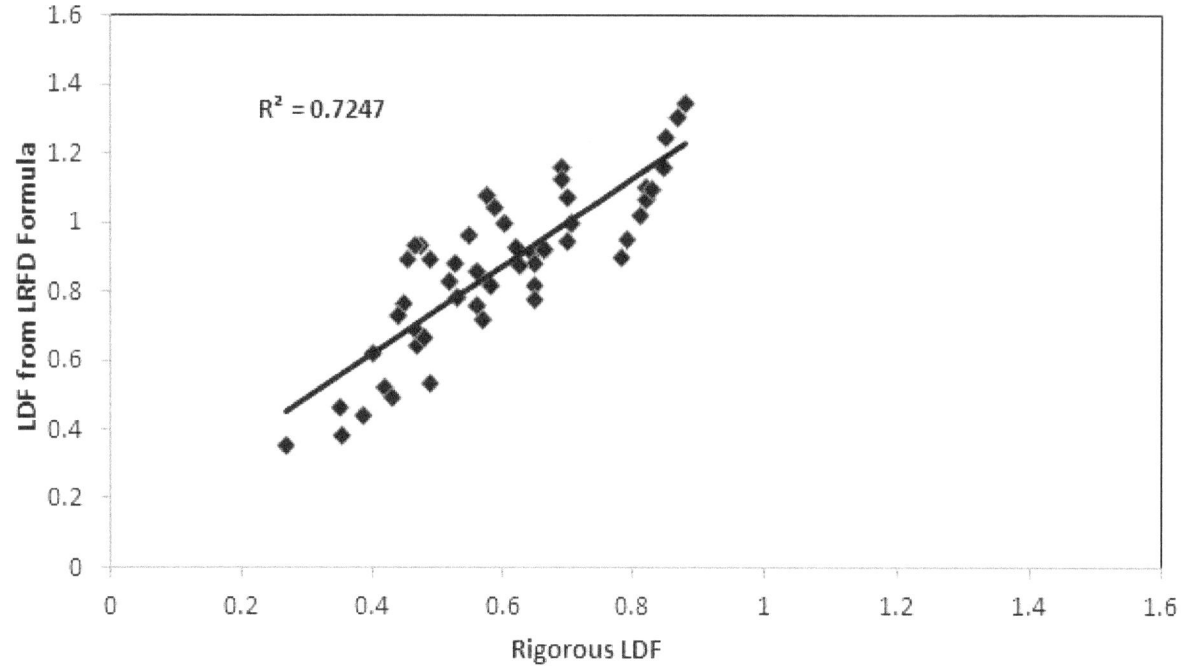

a) Without considering IDs effectiveness

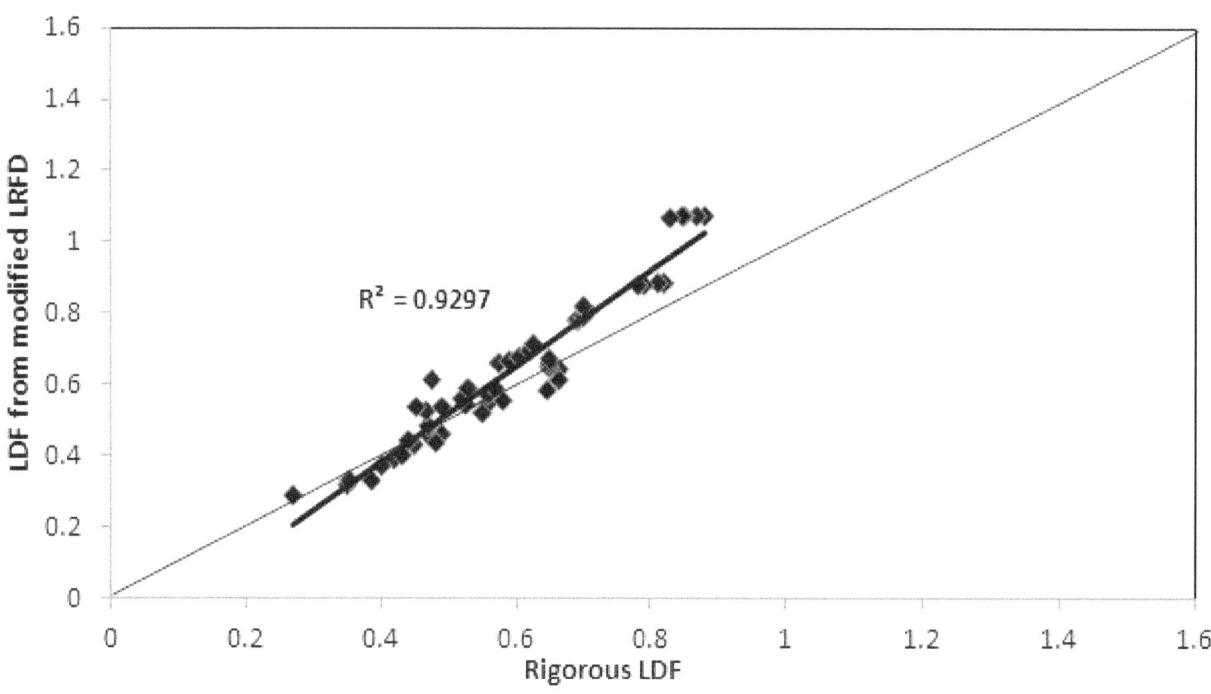

b) With considering IDs effectiveness

Figure 2.47 Live Load Distribution Factor from Modified LRFD Equations vs. Rigorous Finite Element Analysis for Internal Girder

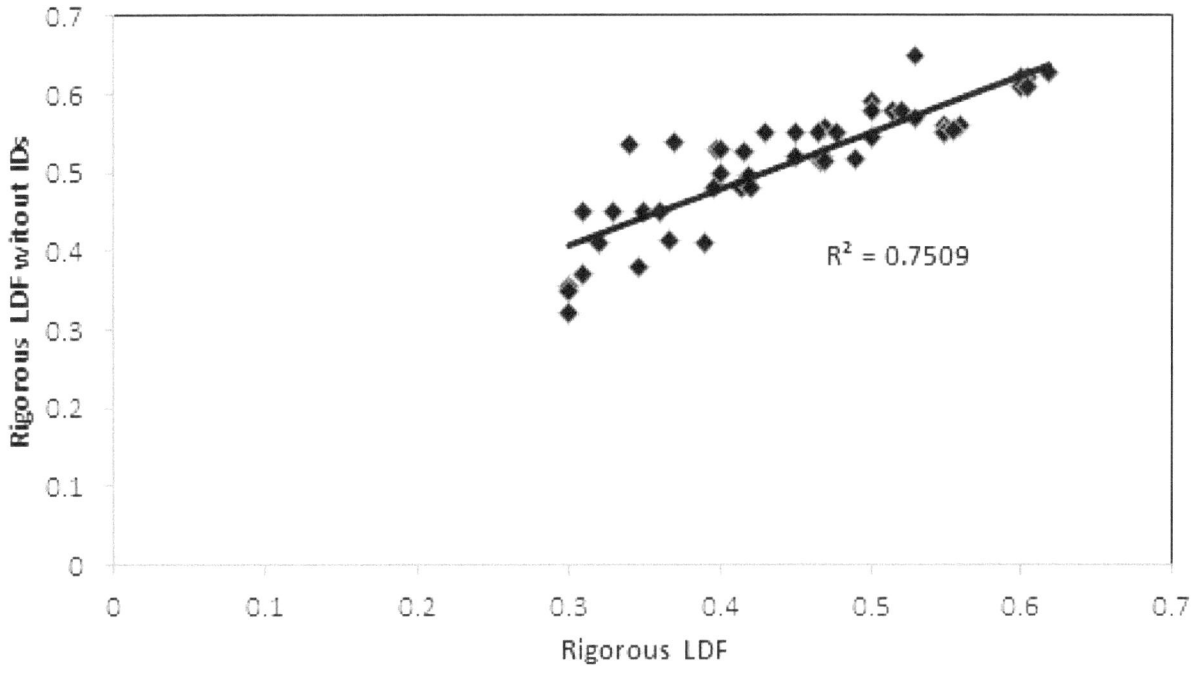

a) Without considering IDs effectiveness

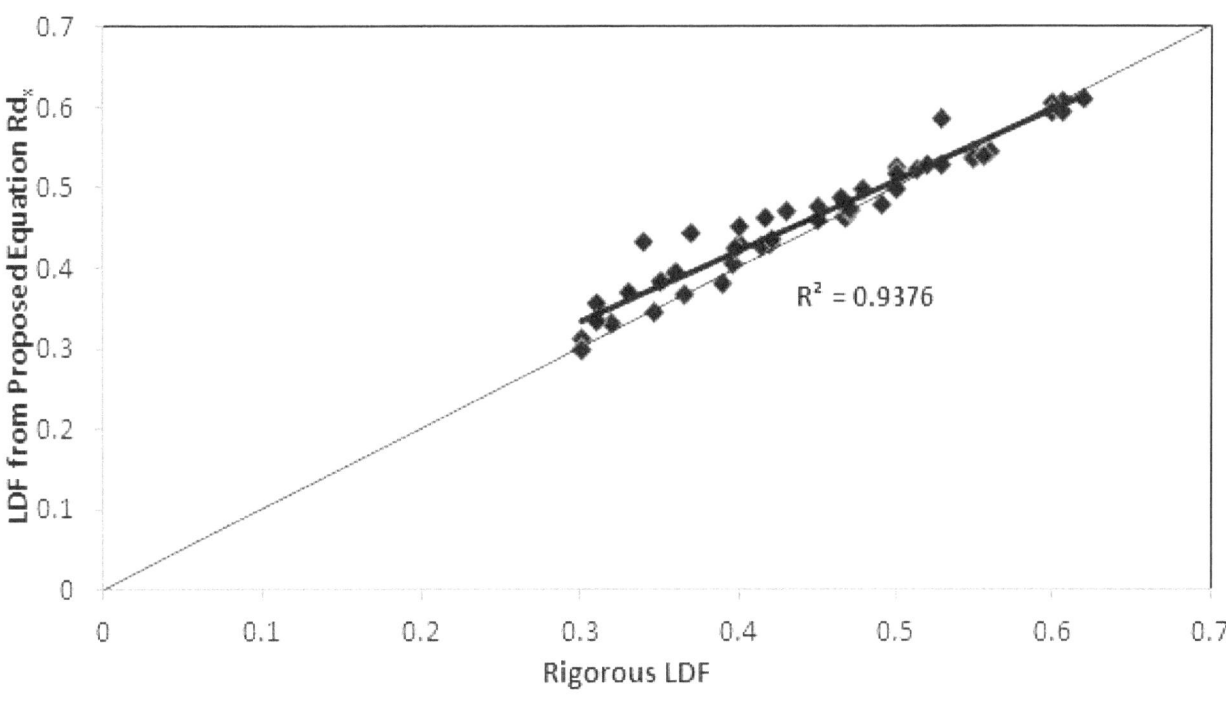

b) With considering the IDs effectiveness

Figure 2.48 Live Load Distribution Factor from Rigorous Analysis and Proposed Equation for external Girder

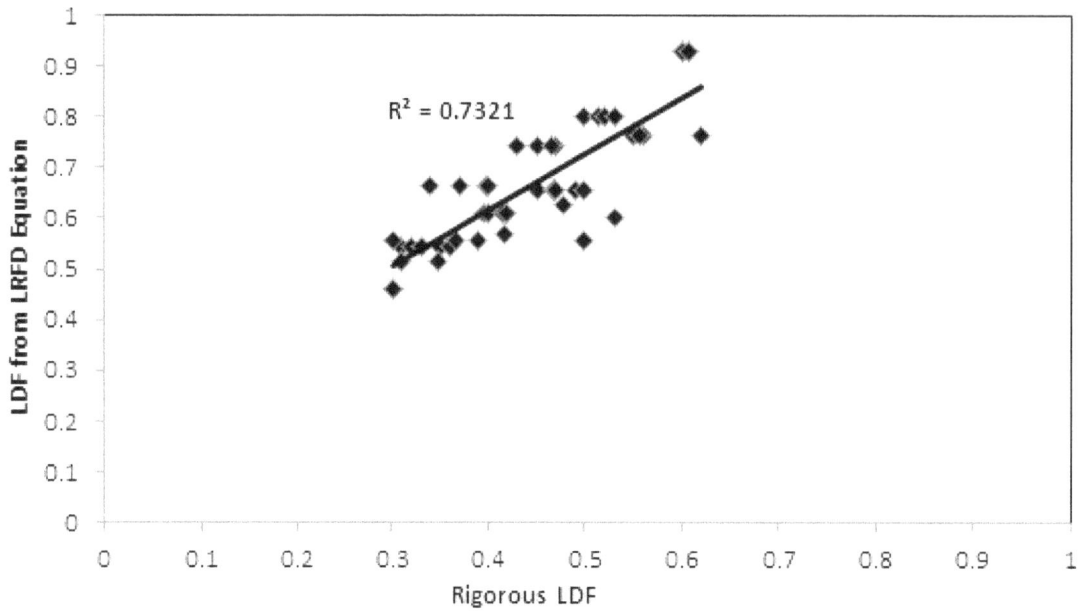

a) Without considering the IDs effectiveness

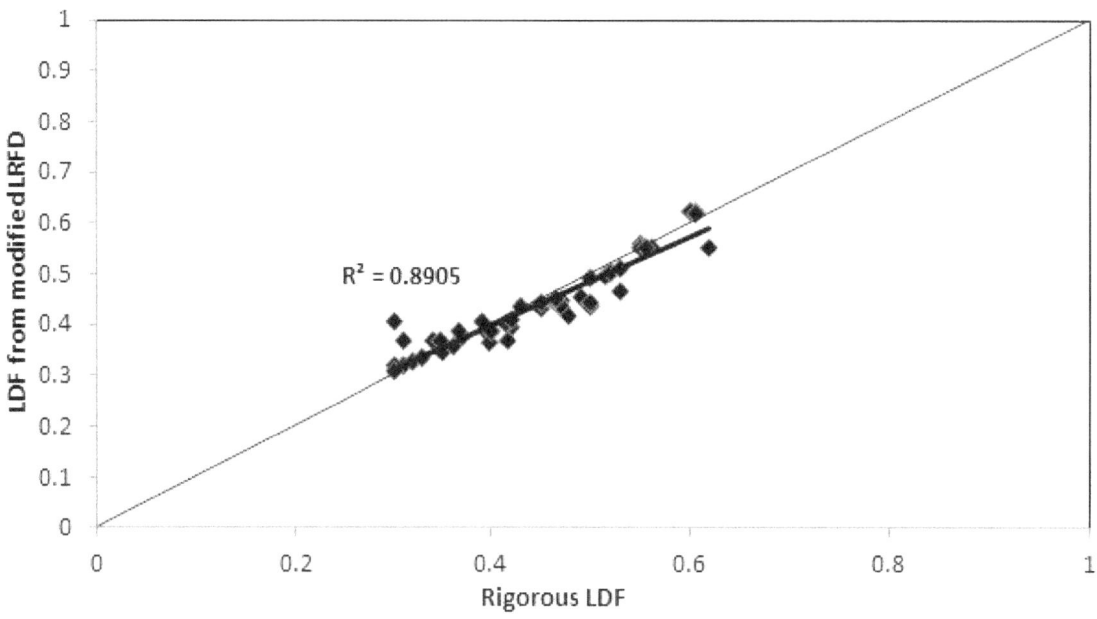

b) With considering the IDs effectiveness

Figure 2.49 Live Load Distribution Factor from Modified LRFD Equation vs. Rigorous Finite Element Analysis for External Girder

In addition, a comparison was performed between distribution factor of bending moment obtained from the finite element analysis and the values retrieved from the "LRFD Based Equation" and "FEA Based Equation" for three box-girders with span length of 60 m, is shown in Figure 2.50 for both external and internal girders. The very slight discrepancy, by up to 3%, between the rigorous method and equations revealed that proposed equations can accurately estimate diaphragms effectiveness in the bending moment distribution factor of bridges.

Table 2.7 Comparative Statistics of Empirical Equations

Distribution factor	Girder	AVG.	SD.	COV.
FEM Based Equations	External	1.038	0.0625	0.0602
	internal	1.009	0.0354	0.0353
LRFD Based Equations	External	0.992	0.0795	0.0766
	internal	1.060	0.1080	0.1018

It was also shown that the effect of the intermediate diaphragms is insignificant in the straight bridges, but rather considerable in the skewed bridges. The AASHTO LRFD (2008) specification were also estimate too conservative results for live load distribution factor of bending moment by average 35% and 31% for external and internal girder, respectively.

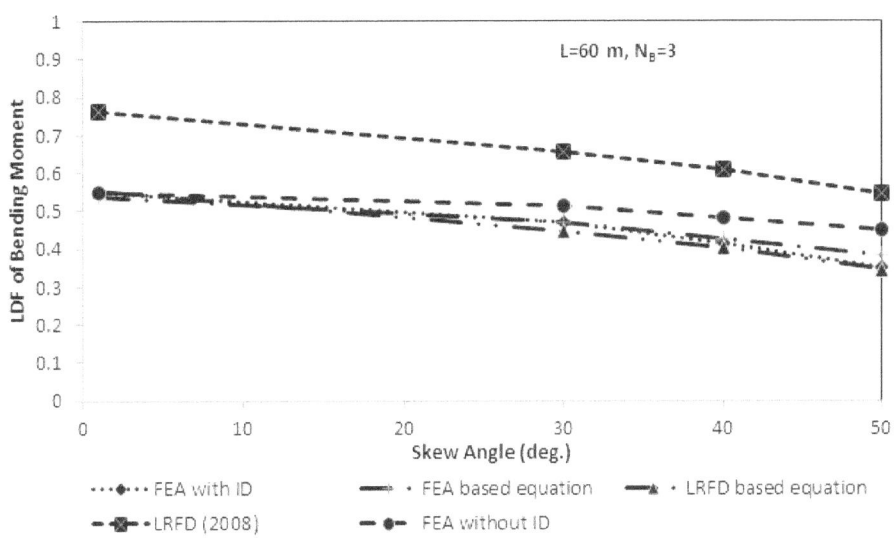

a) External Girder

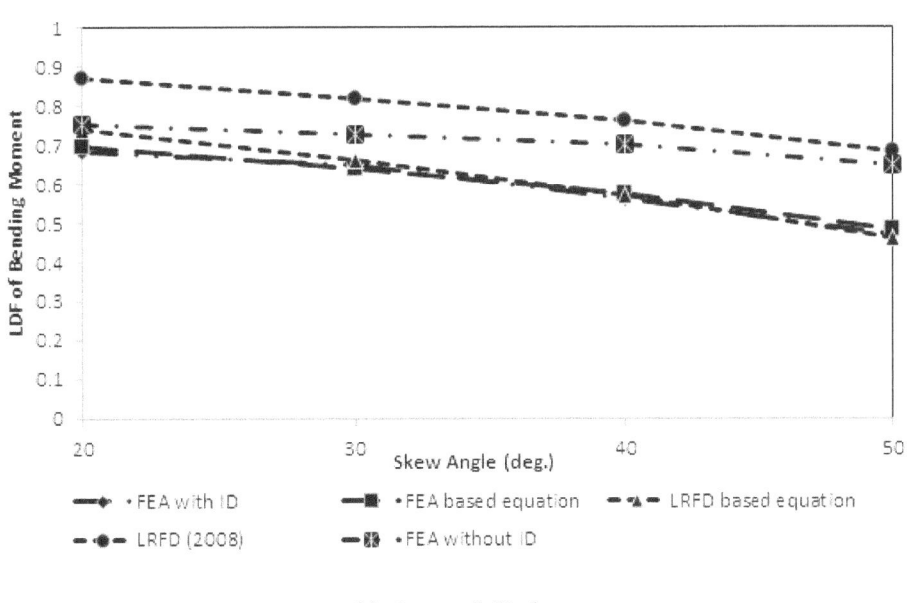

b) Internal Girder

Figure 2.50 Live Load Distribution Factor from Various Method with Intermediate Diaphragms Effectiveness

2.15 Evaluation of Accuracy of Simplified Henry's Method

Even though the AASHTO LRFD method is typically considered to be a more accurate method, accuracy declines when applied to skew and straight bridges of varying lengths and varying member properties. Their complexity and range of applicability often discourage their use by bridge researchers. Hence, in many bridge design procedures, live load distribution factor (LDF) are determined using other simplified methods, such as Henry's EDF method, which only requires the width of roadway and the number of webs in order to calculate the LDF of bridges.

The LDF for bending moment and shear force obtained from simplified Henry's method and finite element technique for bridges with different skew angle are shown in Figure 2.51 and Figure 2.52.

It can be observed that the Simplified Henry's method underestimate the live load distribution factor for bending moment for bridge with small skew angle, and overestimate when skew angle exceed from 45°. It addition, simplified Henry's method obtains so conservative values for live load distribution factor of shear forces. It is because the simplified Henry's method were developed to determine the live load distribution of slab-on-girder bridges, and also does not take into account the effect of main parameters such as skew angle to determine the live load distribution factor.

Hence, to improve the accuracy of simplified Henry's method for computing the distribution of shear force and bending moment, two set of correction factors were deduced. The correction factor equations were derived in the same form of LRFD skew correction factor formula for distribution factor of bending moment.

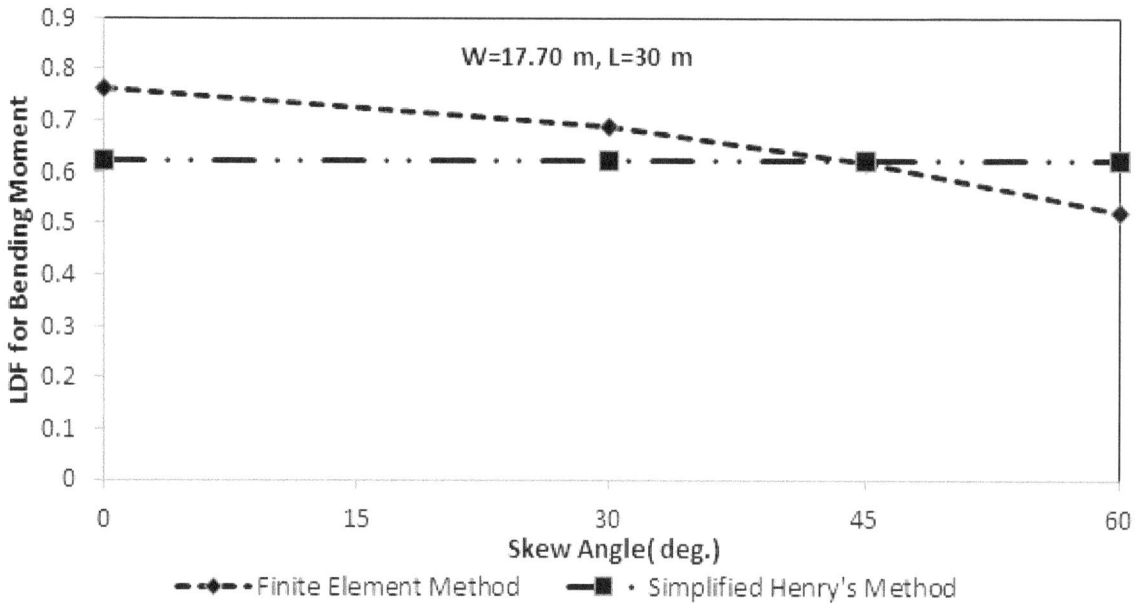

a) Span length of 30 m

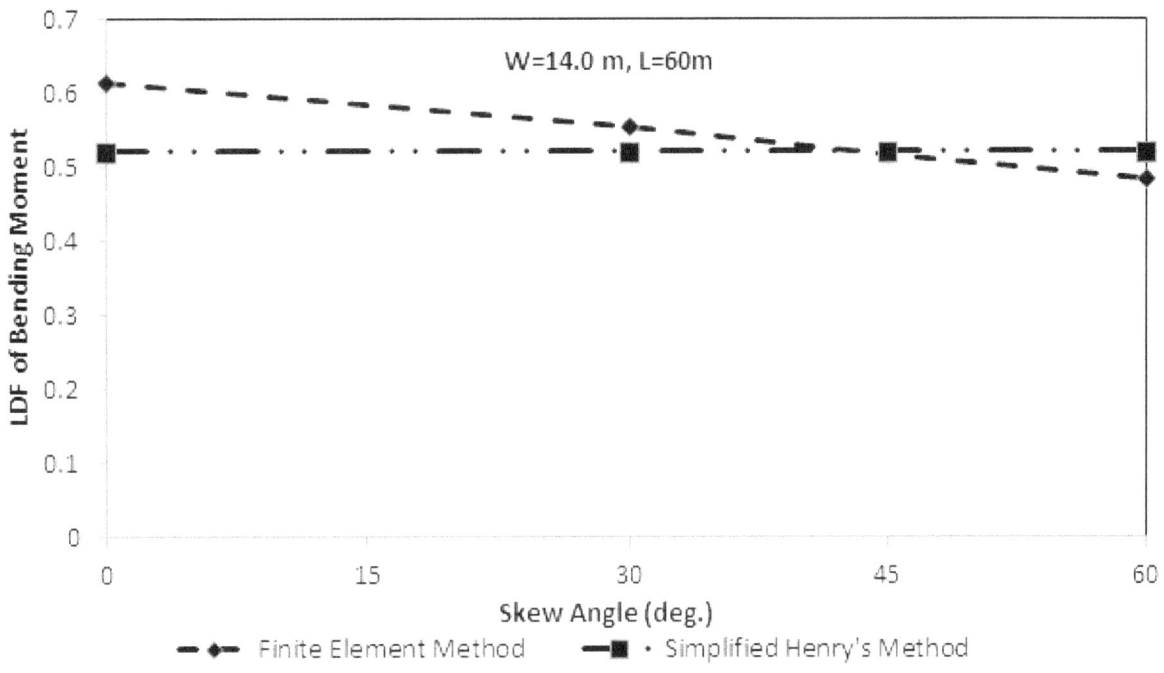

b) Span length of 60 m

Figure 2.51 Bending Moment Distribution Factor from FEM and Simplified Henry's Method

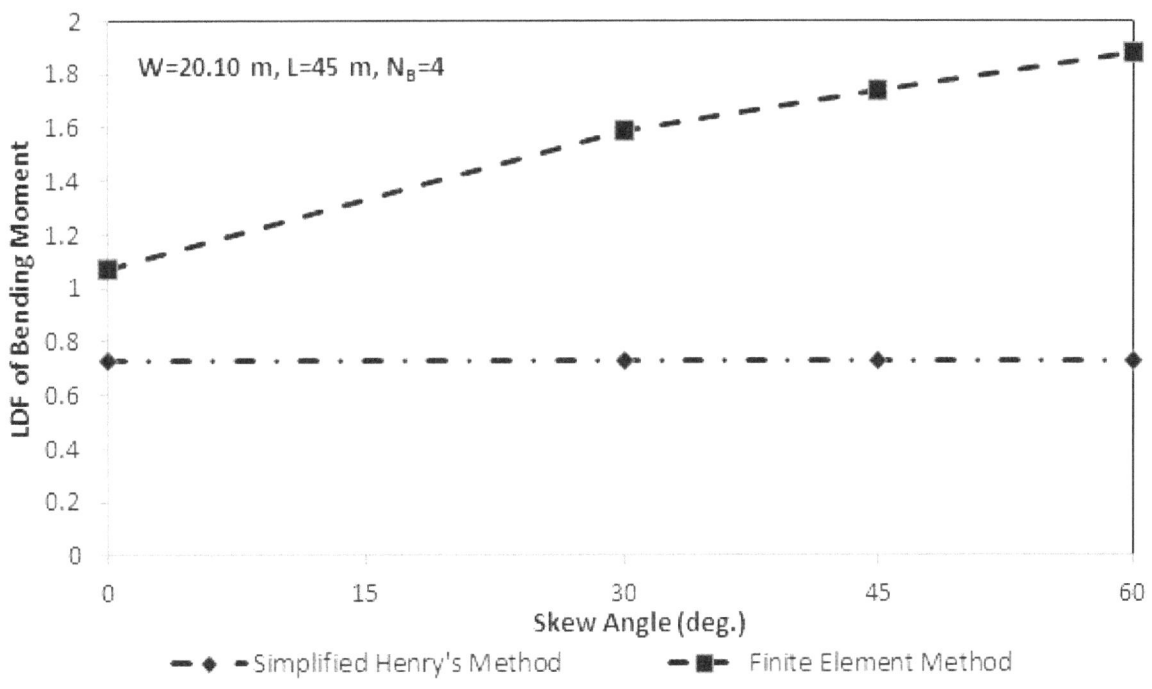

a) Span length of 45 m

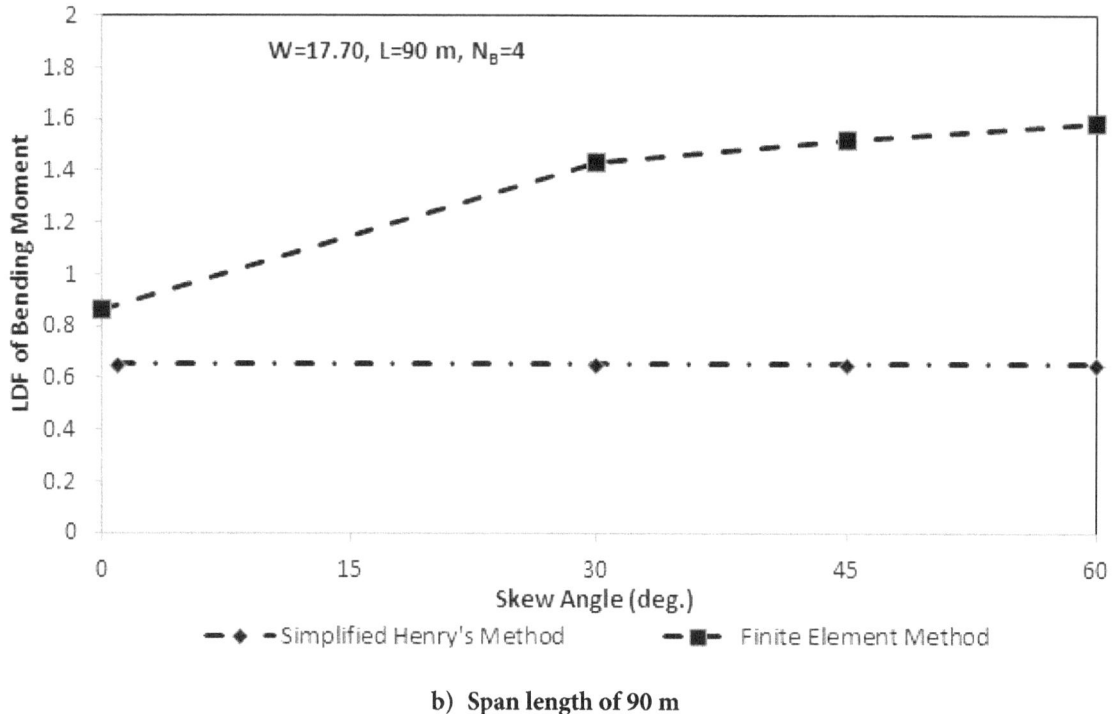

b) **Span length of 90 m**

Figure 2.52 Shear Force Distribution Factor from FEM and Simplified Henry's Method

2.16 Parametric Study on Simplified Henry's Method

Due to the skew correction factor formula for multicell box-girder bridges in the current AASHTO LRFD (2008) specifications include the term, L/d (Span length vs. deck depth), and skew angle. This term was kept in the proposed skew correction factor equations for simplified Henry's method. Then the effect of L/d, skew angle, and girder spacing, representation of bridge width on skew correction factor for live load distribution factor of bending moment and shear force were investigated.

The skew correction factor (SCF) was obtained as ratio of live load distribution factor of skew bridge to live load distribution factor of straight (θ=0) bridge ($LDF_{skew}/LDF_{straight}$). The prototype bridge presented before was used for this section to consider the effect of ratio L/d in developing the correction factor equation for simplified Henry's method.

2.16.1 Effect of Span-to-Depth Ratio

Figure 2.52 shows the effect of L/d on skew correction factor of live load distribution factor of bending moment obtained from finite element analysis for a bridge with total width of 17.70 m from Table 2.6. The skew angle changed from 30° to 60° in 15° increment.

The skew correction factor for live load bending moment increased as L/d increased. As can be observed in Figure 2.53, the skew correction factor for live load bending moment was increased by up to 23.0%, as L/d changed from 1.58 to 2.00 at the skew angle of 60°.

The effect of span-to-depth ratio (L/d) on skew correction factor SCF of live load distribution for shear force, for a bridge with total width of 17.70 m, is shown in Figure 6.54. It can be observed that relationship between L/d and correction factor of shear followed the same manner as that for bending moment so that with increase the skew the ratio, the skew correction factor increased. This increase was almost 60% for bridge with skew angle of 30°. For skew bridge higher than 45° the effect of skew angle on this relationship was marginal. Thus, the effect of L/d should be considered in developing the proposed equations for modification of simplified Henry's methods.

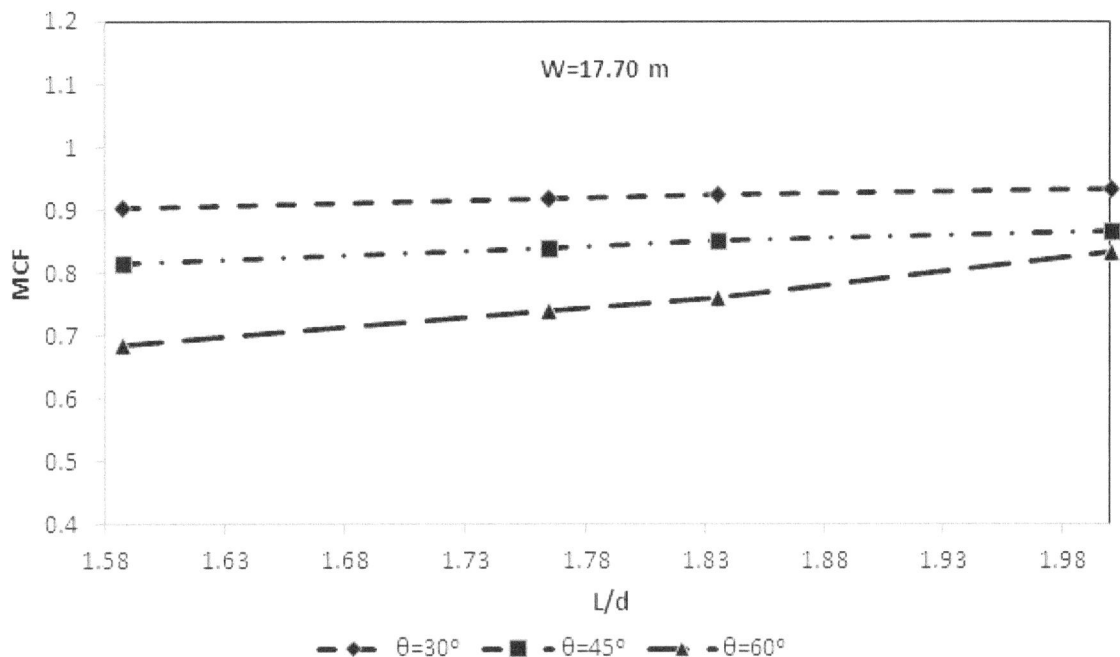

Figure 2.53 Moment Skew Correction Factor vs. Span-to-Depth Ratio

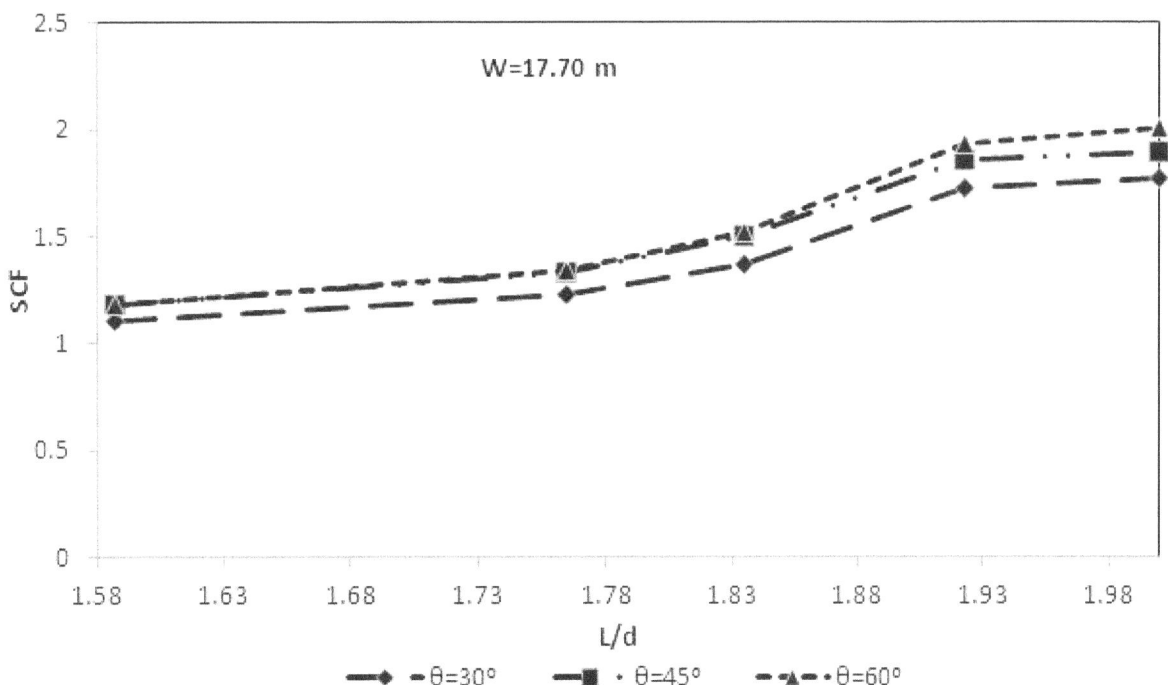

Figure 2.54 Shear Skew Correction Factor vs. Span-to-Depth Ratio

2.16.2 Effect of Skew Angle

The skew angle in multicell box-girder bridge is another main parameter considered in the proposed skew correction factor equations. Figure 2.55 and Figure 2.56 indicate the influence of skew angles on skew correction factor for live load distribution of bending moment and shear forces, respectively. The span length of 30 m, 60 m and 90 m were randomly chosen to present the results.

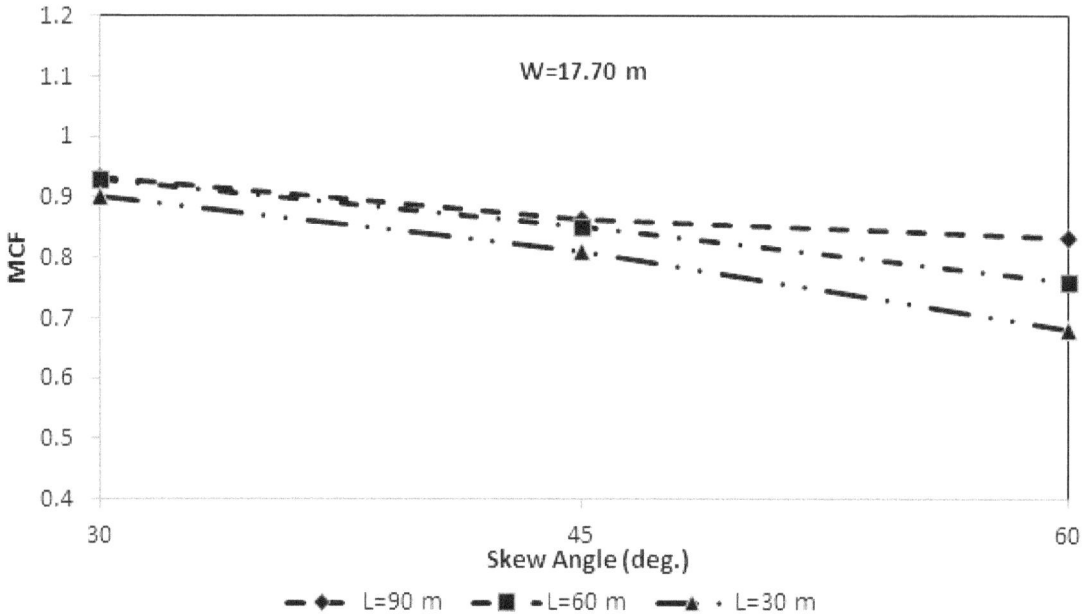

Figure 2.55 Moment Skew Correction Factor vs. Skew Angle

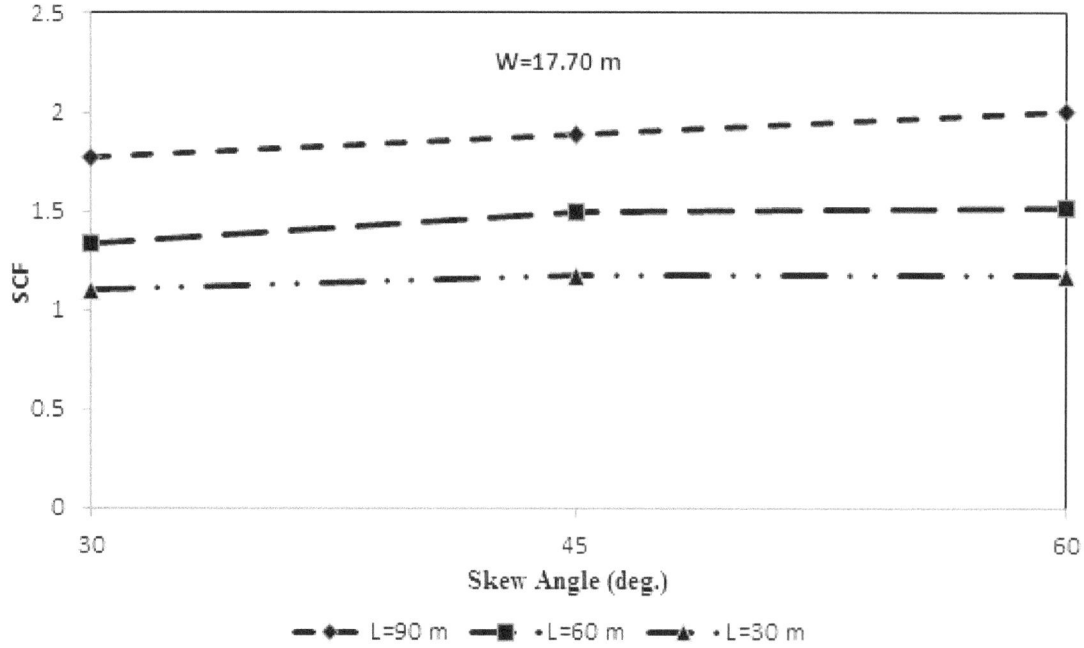

Figure 2.56 Shear Skew Correction Factor vs. Skew Angle

It observed that the skew correction factor SCF of live load distribution of bending moment decreased when skew angle increased. The skew correction factor declined by 25%, as skew angle changed from 30° to 60°.

From Figure 6.56, it can be seen that the skew correction factor of shear force increased, as the skew angle increased. The factor increased by almost 14% when skew angle varied from 30° to 60°. Thus, the effect of skew angle was more significant for skew correction factor of live load distribution factor of bending moment.

2.17 Correction Factor Equations for Simplified Henry's Method

In this section the derivation of the equations to determine the skew correction factor for shear force (SCF) and bending moment (MCF) for single span and continuous multicell box-girder bridges were presented, separately.

The equations were obtained in the same form of the skew correction factor formulas for AASHTO LRFD (2008) specifications. The parametric study indicated that the equations should been presented in term of skew angle θ, girder spacing S, and span-to-depth ratio L/d.

The adopted method from the NCHRP 12-26 project conducted by Zokaie et al. (1993) was used to develop the proposed correction factor equations. The results of finite element analysis on the 80 prototype bridge presented in Table 2.6 were employed for this study. The following form was selected to develop the correction equations;

$$SCF = 1 + a \times \left(1 + b.\left(\frac{L}{d}\right)\right) \times \left(f\left(tag\theta\right)\right) \times \left(f\left(s\right)\right) \qquad (2.26)$$

Based on the method described in Chapter III, the following equations were obtained for correction factor of moment (MCF) and shear (SCF) of single span multicell box-girder bridges, respectively.

$$MCF = 1.04 + \left(1 + 0.15 \times \frac{L}{d}\right)\left(0.10 - 0.11 \times \tan\theta\right)\left(0.70 + 0.036 \times S\right) \qquad (2.27)$$

$$SCF = 1 + \left(1.15 + 0.24 \times \frac{L}{d}\right)\left(0.8 \times \tan\theta^{0.27}\right)\left(0.71 \times S^{0.17}\right) \qquad (2.28)$$

A closer look revealed that the effect of the third term of the equations is insignificant (5% and 7.1%, respectively), so it can be underestimated. Using a same method as described for single span bridges, the following equations were proposed to determine the correction factor for shear SCF and moment MCF of simplified Henry's method for continuous bridges. The main parameters were those selected for single span bridges.

$$MCF = 1.05 + \left(0.25 + 0.0107 \times \frac{L}{d}\right)\left(0.223 - 0.13 \times \tan\theta\right) \qquad (2.29)$$

$$SCF = 1 + \left(0.427 + 0.017 \times \frac{L}{d}\right)\left(\tan\theta\right)^{0.23} \times A \qquad (2.30)$$

where

$$A = 0.29 \times S^2 + 1.84 \times S - 1.675 \qquad (2.31)$$

It can be observed that the Eq. (2.28) and Eq. (2.29) are only as function of span-to-depth ratio and skew angle, and the effect of girder spacing did not considered in developing the equation.

2.18 Verification of Proposed Equations

The regression analysis was employed to investigate the proposed correction factor of Eq. (6.26) through Eq. (6.29). As indicated in Figure 6.57 and Figure 6.58, the relationship between live load distribution factor through rigorous analysis and simplified Henry's method, with and without multiplying the proposed skew correction equations were plotted.

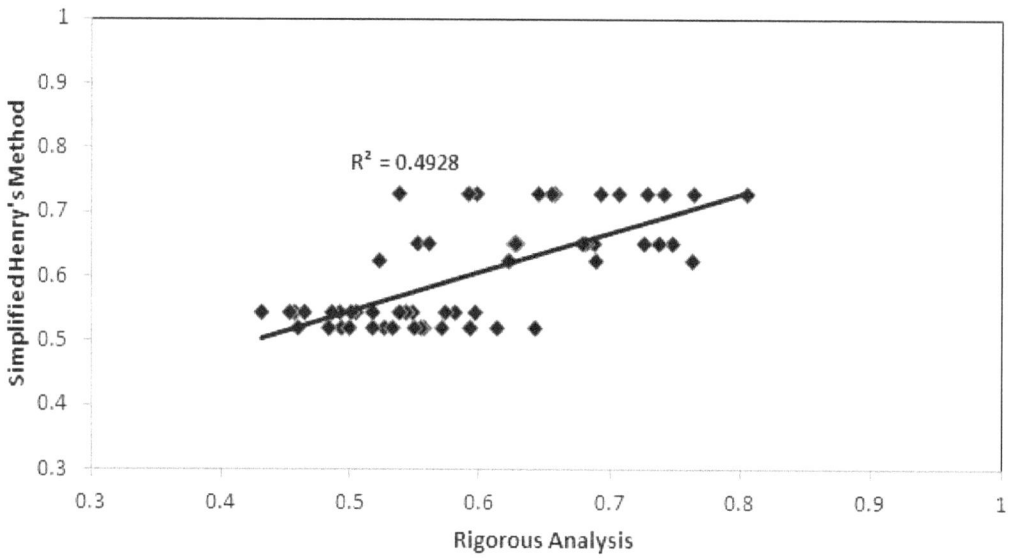

a) **Without applying correction factor**

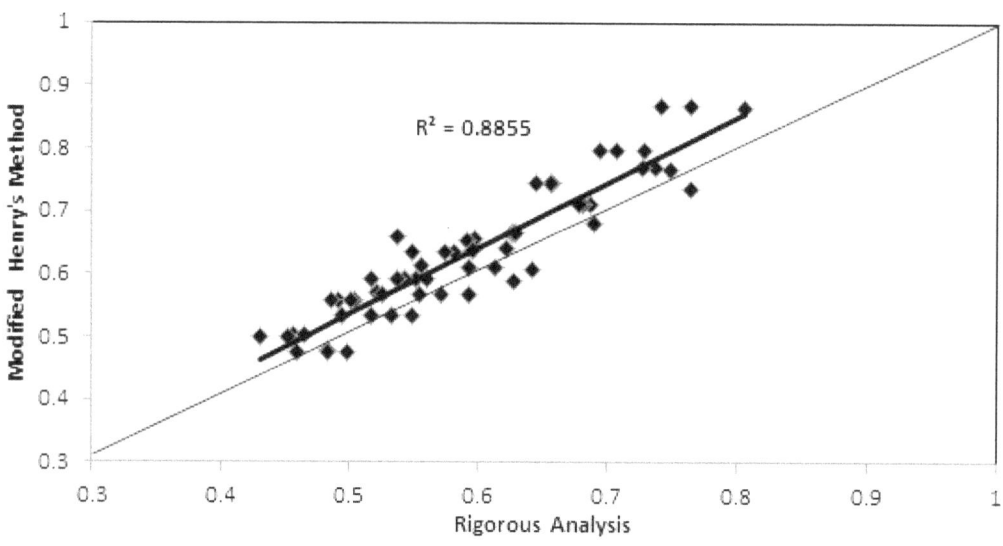

b) **With applying correction factor**

Figure 2.57 Simplified Henry's Method vs. Rigorous Live Load distribution Factor for Bending Moment of Single Span Bridges

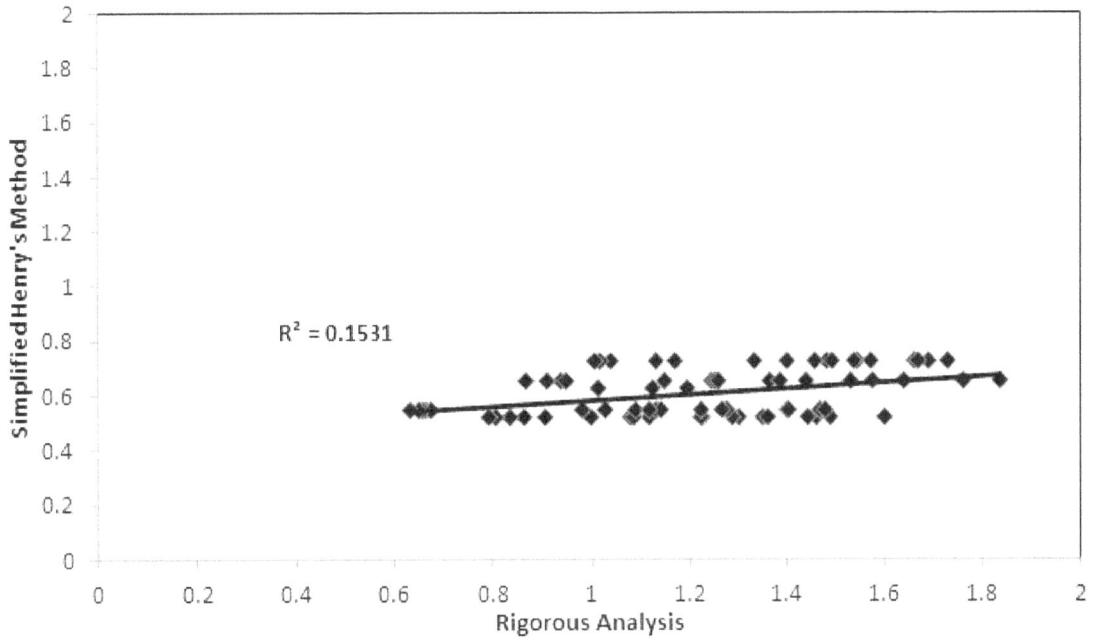

a) Without applying correction factor

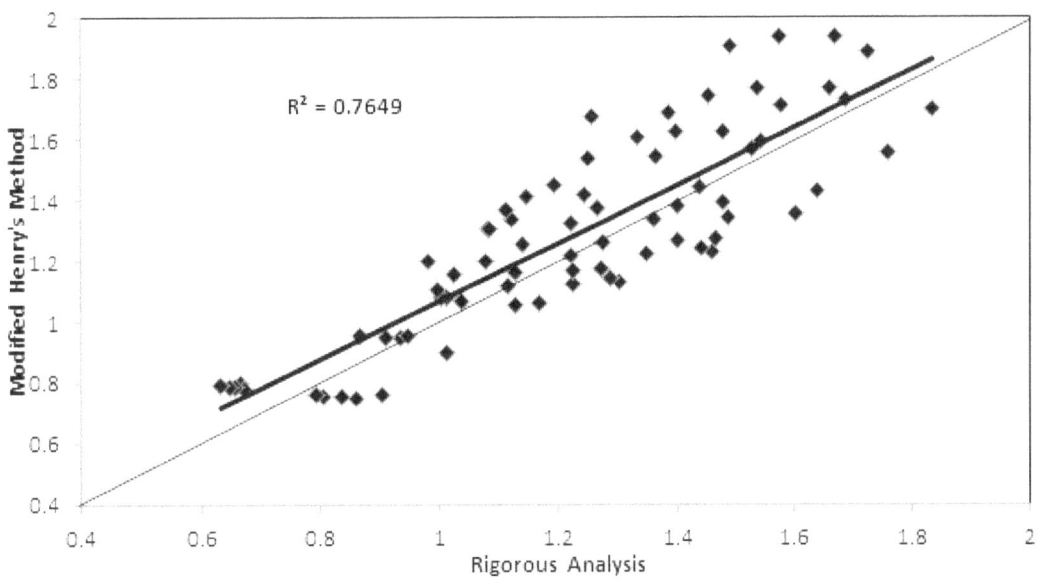

b) With applying correction factor

Figure 2.58 Simplified Henry's Method vs. Rigorous Live Load Distribution Factor for Shear Force of Single Span Bridges

It was observed that the proposed equations reduced the scattering of results so that the coefficients of determinations R^2 of live load distribution factor of bending moment changed from 0.4928 to 0.8855 and also varied from 0.1531 to 0.7649 for the shear force, which indicates a slight deviation compared with rigorous analysis. Although the coefficients of determinations R^2 for live load distribution factor of shear force was not so large, the slightly higher to unity average and small standard deviation presented in Table 2.7 revealed that the above proposed equations can be applied economically and conservatively in design of multicell box-girder bridges.

The relationship between finite element results and simplified Henry's method with applying correction factor are presented in Figure 2.59 for two span continuous bridges. The variation of live load distribution factor before applying correction factor almost followed same trend as presented in Figure 2.57 and Figure 2.58.

In Table 2.8, the average AVG., standard deviation ST.D and coefficients of variation COV, for live load distribution factor of proposed correction factor equation were presented. The maximum standard deviation were 0.1267 and 0.08410 for distribution of bending moment and shear force, respectively, which means obtained results were so close to mean. The maximum coefficient of variation of the results from proposed equations to finite element analysis for shear and bending moment were 0.1206 and 0.0808, respectively, which were excellent. It should be mentioned that an advantage of simplified Henry's method is that it did not limit for design of bridge with range of applicability.

In Figure 2.60 and Figure 2.61 the live load distribution factor obtained from various method were calculated for bending moment and shear force of a bridge with 60 m span. The live load distribution factors from the AASHTO (2002) standard, AASHTO LRFD (2008) specifications, Simplified Henry's methods, with and without modification factor, and finite element analysis are presented.

As can be observed from Figure 2.59, LRFD formulas estimated conservative value for distribution factor of bending moment for internal girder, and predicted unsafe results for external girders. The traditional AASHTO standard equation which is only a function of girder spacing obtained values that were very close to Simplified Henry's method in this case. The results of the modified simplified Henry's method (proposed equations) were up to 8% higher than rigorous data.

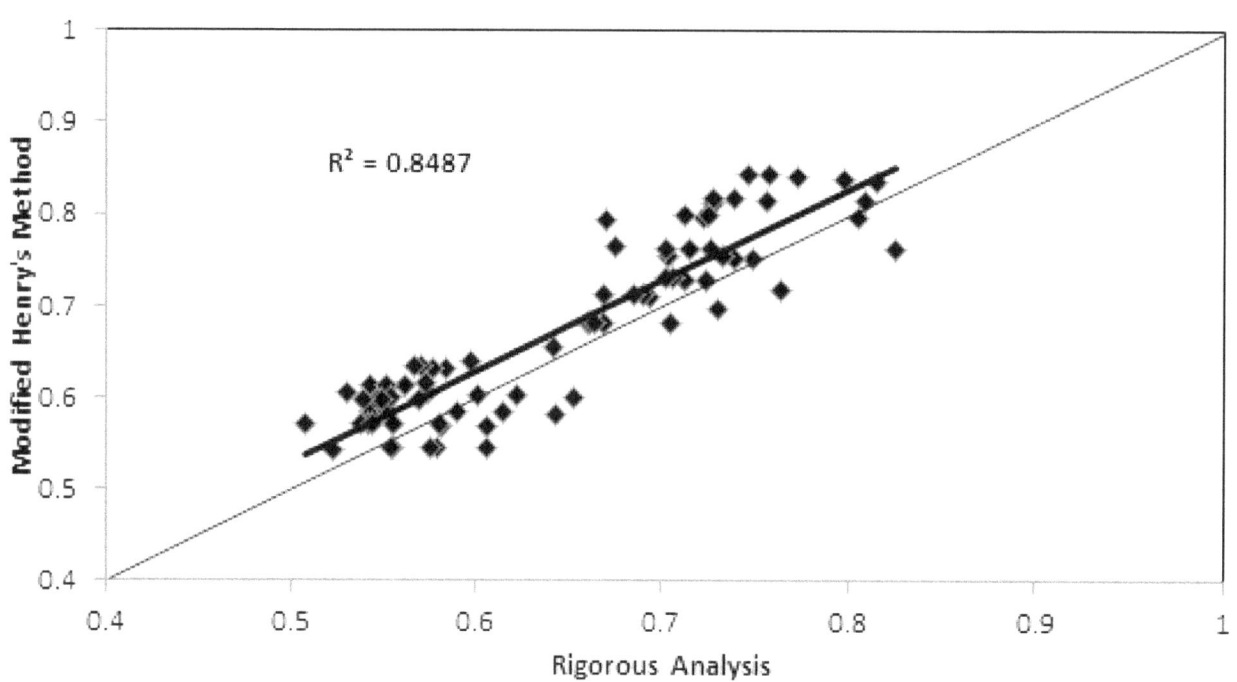

a) Live load distribution of Bending moment

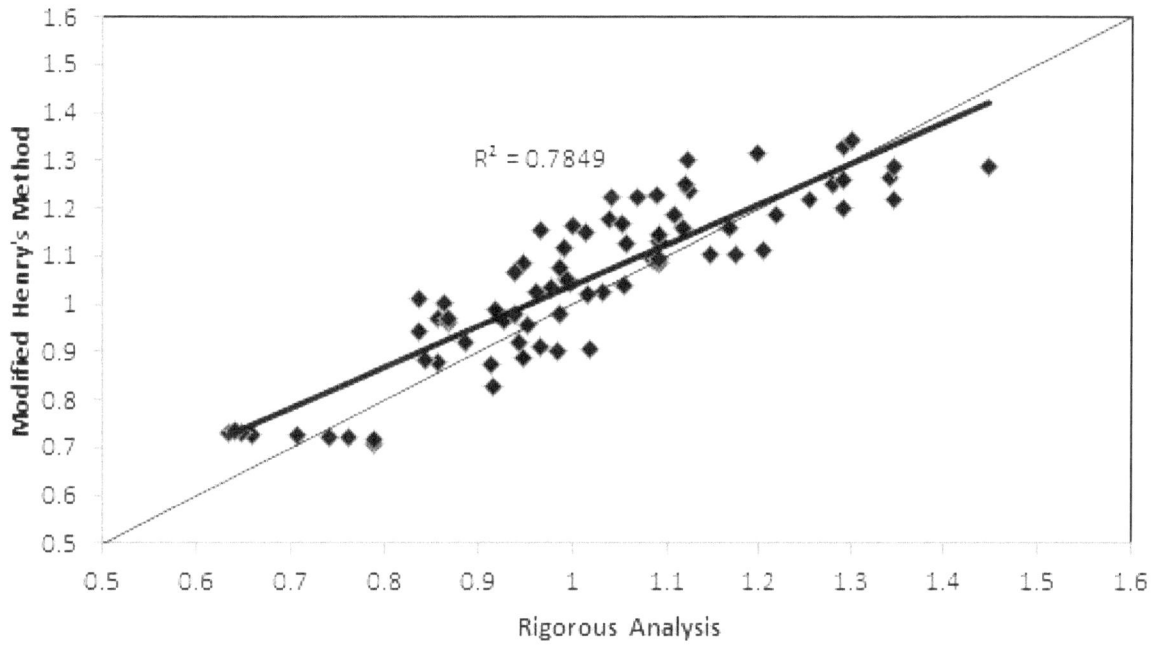

b) Live Load Distribution Factor of Shear force

Figure 2.59 Modified Simplified Henry's Method vs. Rigorous Live Load Distribution Factor for Continuous Two Span Bridges

Table 2.8 Comparative Statistics of proposed equations

LDF	AVG. (Equation/Rigorous)	ST.D	COV.
Bending Moment-one span	1.070	0.1267	0.12060
Shear Force-one span	1.052	0.06336	0.05925
Bending Moment-two span	1.044	0.06210	0.06010
Shear Force-two span	1.040	0.08410	0.08082

From Figure 2.61, it can be seen that AASHTO (2002) standard and simplified Henry's method underestimated the live load distribution of shear force. LRFD formula obtain relatively acceptable value for internal girder of multicell box-girder bridge, however, it obtained conservative results for bridge with highly skew angle. The LRFD formula predicted very unsafe value for external girder of bridges. The results obtained from proposed equation for modified simplified Henry's method were almost 5% higher than those from rigorous analysis. The close agreement between finite element values and modified Henry's method revealed that the proposed equations can be used in bridge design procedures.

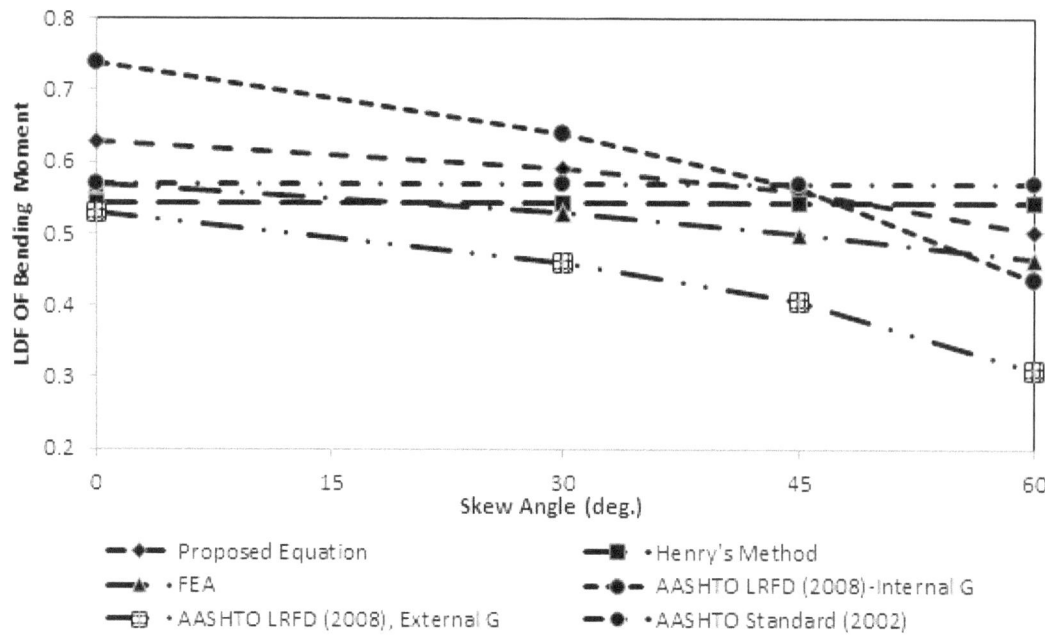

Figure 2.60 Live Load Distribution of Bending Moment from Various Analytical Methods

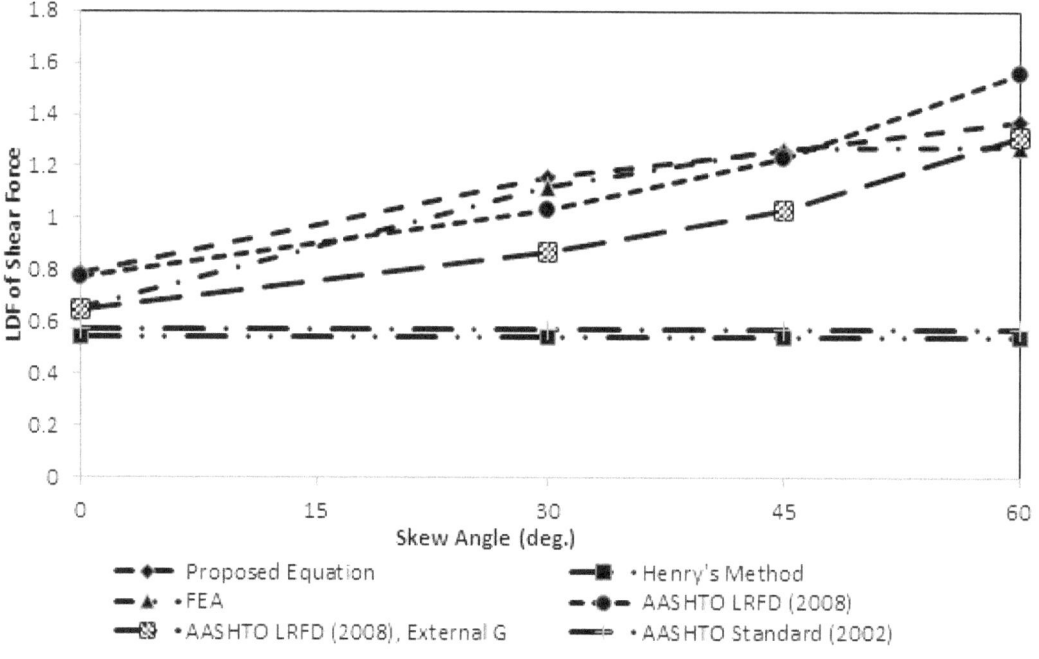

Figure 2.61 Live Load Distribution of Shear from Various Analytical Methods

2.19 Secondary Bending Moment

Generally, the effect of torsional moment in the regions where the maximum bending moment is small can be underestimated. In skewed decks, torsion includes transverse and negative flexural bending that must be considered in bridge design. Although the advanced finite element program is able to take these effects into consideration they are often very time-consuming. Meanwhile, most specifications (AASHTO LRFD and AASHTO specifications) and simplified methods (Henry's method) are unable to determine the secondary moment of bridges.

The grillage analysis which is a fast and reliable method requires a post-processing to determine the torsional moment of the superstructure that is partly confusing. To find a solution to this problem, the results of the FEA from a parametric study on prototype bridges were employed to obtain the ratio of the secondary moments (positive and negative torsion) to the maximum bending moment, as shown in Figure 2.62.

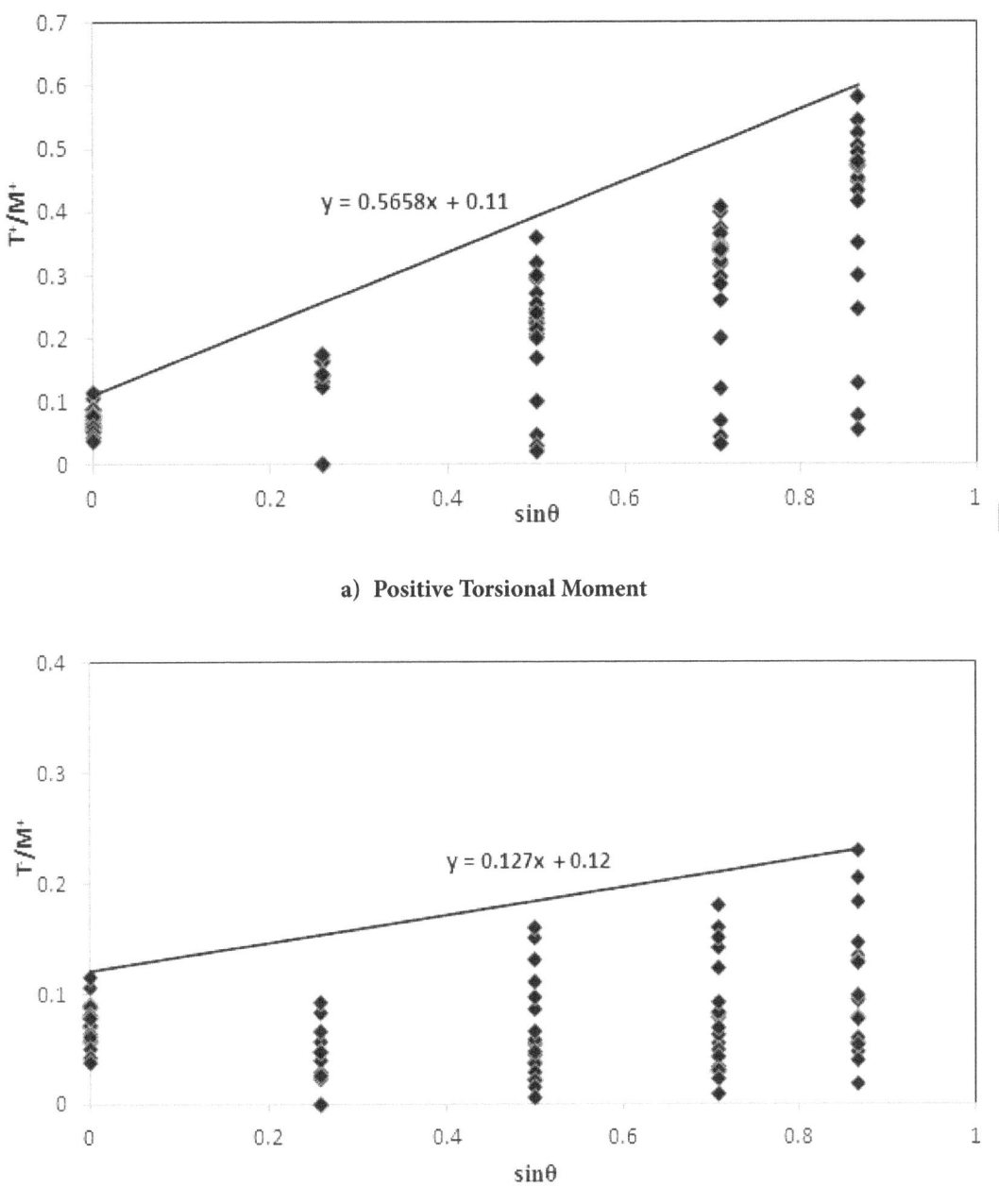

a) Positive Torsional Moment

b) Negative Torsional Moment

Figure 2.62 The Ratio of Secondary Moment and Positive Bending Moment

Based on the upper bound envelopes of values of the ratio each secondary moment to corresponding maximum bending moment, to keep a safe level, the following expression was deduced;

– Positive torsional moment:

$$T^+ = M^+ \left(0.566 \times \sin\theta + 0.11\right) \qquad\qquad 0 \leq \theta \leq 60 \qquad (2.31)$$

– Negative torsional moment:

$$T^- = M^+ \left(0.127 \times \sin\theta + 0.120\right) \qquad\qquad 0 \leq \theta \leq 60 \qquad (2.32)$$

2.20 Live Load Distribution of Reaction at Piers

Live load distribution factors for bending moment and shear force are calculated using the AASHTO LRFD (2008) Specifications. The current codes do not recommended any formulas to determine the distribution of reaction. Hence, reaction distribution factor of skew bridges was calculated either by the shear skew correction factor (SCF) defined by AASHTO LRFD (2008) or by using LRFD formulas without considering the effect of skewness. Although a significant difference between live load distribution factor of shear and reaction was observed (Zheng 2008).

In Figure 2.63 and Figure 2.64 the live load distribution factor of shear force and reaction for bridge with skew angle ranged from 0 to 60° and with bridge width of 11.60 m and 17.70 m were compared to identify the difference between the values of shear and reaction distribution factor for skew bridges. For all skew bridge subjected to applied loading condition used in this research, the maximum reactions were obtained in the external girder at the acute corner of skew bridges.

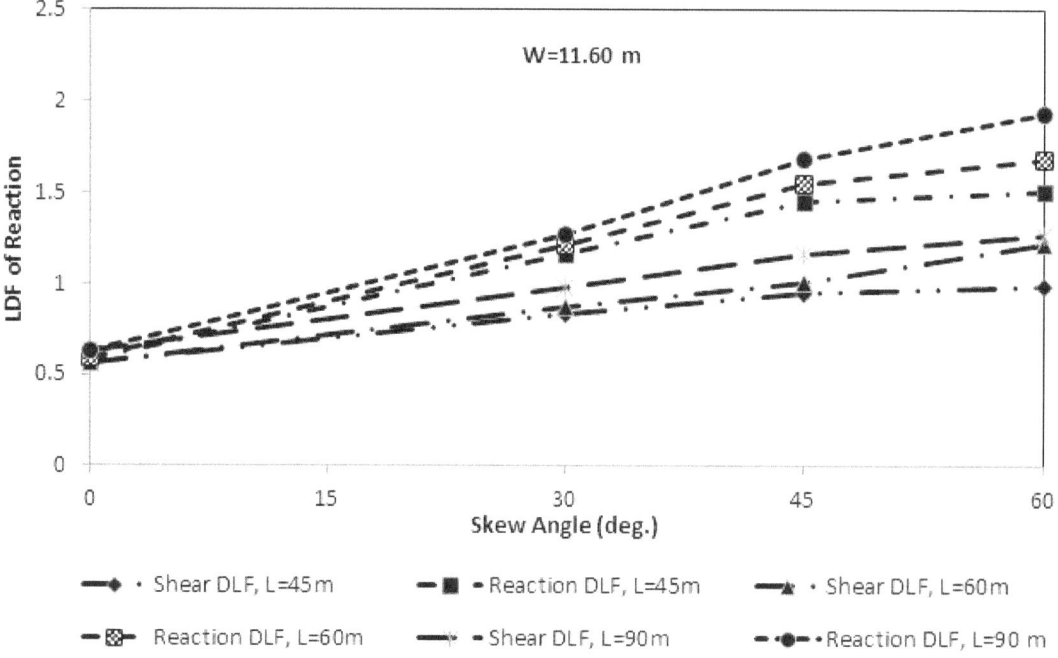

Figure 2.63 Live Load Distribution Factor for Shear and Reaction for Bridge with Different Span Length

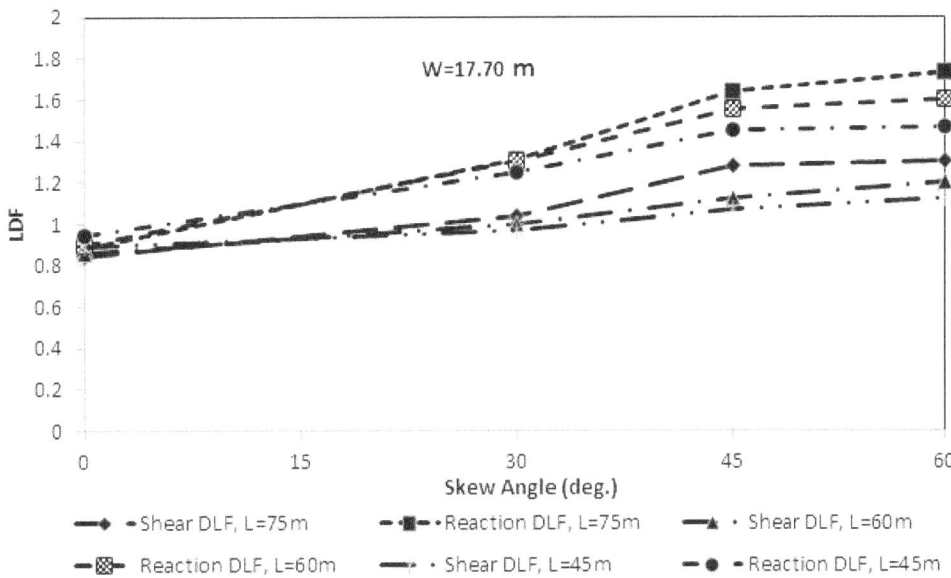

Figure 2.64 Live Load Distribution Factor for Shear and Reaction for Bridge with Different Span Length

It can be observed from Figure 2.63 and Figure 2.64 that the difference between these two values increased when skew angle increased. The distribution factor of shear and reaction are almost same when in straight bridges, so LRFD formula can be used to obtain this value. Figure 2.65 and Figure 2.66 compared the live load distribution factor from various analytical methods for bridge with span length of 60 m and 75 m.

As already described, the AASHTO LRFD (2008) proposed skew correction factor expressions to consider the effect of skewness on live load distribution factor of shear. On the other side, the lever rule technique was developed based on the static equilibrium of a girder with simply supported, without considering the skew effect. In this section, the live load distribution factor for each bridges were calculated from LRFD formula, AASHTO standard, Henry's method, with and without considering the skew correction factor of LRFD, finite element analysis, Revel Rule approach. Modified Revel Rule approach was also determine with multiplying the LRFD skew correction factor in obtaining the results.

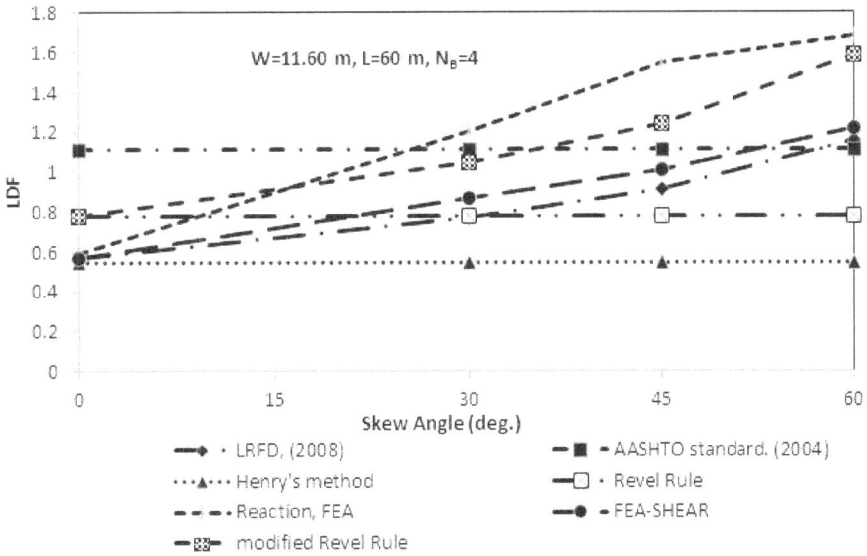

Figure 2.65 Live Load Distribution Factor of Reaction at Piers from
Various Analytical Methods for Bridge with 60 m Span Length

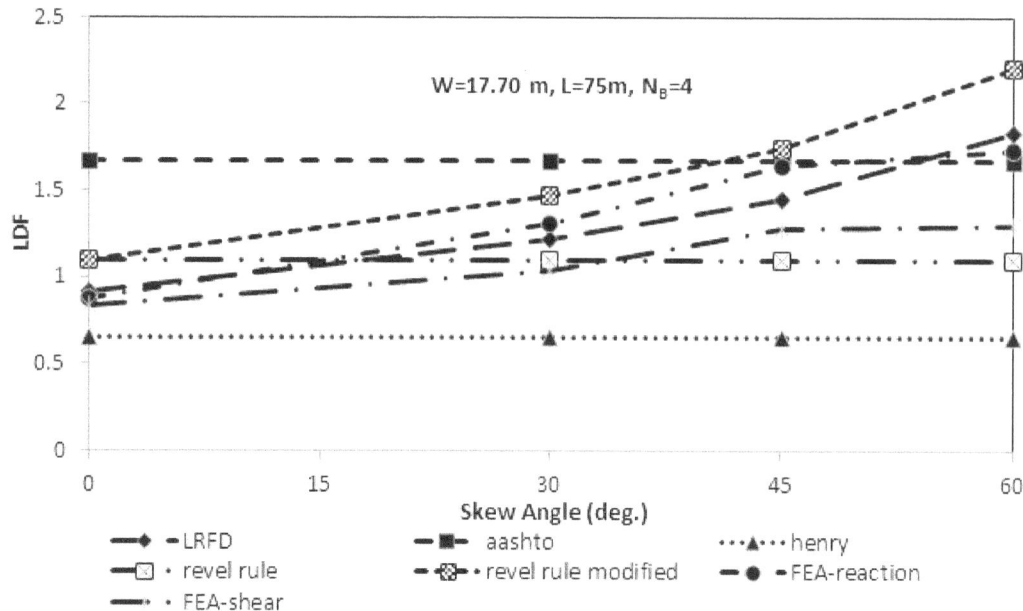

Figure 2.66 Live Load Distribution Factor of Reaction at Piers from Various Analytical Methods for Bridge with 75 m Span Length

It can be observed that LRFD formulas underestimated the maximum reaction distribution factor which obtained at the external girder at the pier of bridges. The same finding was obtained for simplified Henry's method and Revel Rule approach. On the other hand, the results of modified Revel Rule presented more reasonable value for live load distribution factor of reaction at the piers; however, the modified Revel Rule method did not obtain safe value for all bridges. Thus, in the primary stage of design, modified Revel Rule can be used to predict a reasonable value for live load distribution factor of skew multicell box-girder bridges. But future study is required to develop the skew correction factors for obtaining the accurate value for distribution of reaction at the piers.

2.21 Live Load Distribution of Torsional Moment

Previous analytical studies indicate that currently available specifications are unable to consider the effect of the twisting moment (torsional moment) on bridge actions. In straight bridges the effect of torsion is negligible and the transverse reinforced design is governed by other requirements. However, in the case of skewed bridges the effect of the twisting moment should be considered. Therefore, an in-depth study was performed on 80 concrete multicell box-girder bridges with skew angles ranging from 0 to 60°.

2.21.1 Torsional Stiffness of Multicell Cross Section

Sennah & Kennedy (1999) indicated that the stiffness of a composite cellular structure depends on the thickness of the steel plates and deduced several expressions for torsional stiffness (J) for this type of bridge.

To develop an expression for the torsional stiffness of concrete multicell box-girder cross sections, the effect of the number of boxes (N_B) on torsional stiffness was investigated. There are two ways to determine torsional stiffness (J) of multicell box-girder bridges, including the precise method (simultaneous Equations method) and the approximation method, in which the effect of the internal web is neglected in computing the torsional stiffness. Figure 2.67 shows the variation of torsion-to-flexural stiffness versus the number of boxes (N_B).

It became clear that with neglecting the internal webs in computing the torsional stiffness, the ratio increased by only 5%, which indicated that internal webs have an insignificant effect on the distribution of shear flow over the cross section. The approximation method calculated torsional stiffness up to 7% higher than that from the exact method. Therefore, the following simplified expression obtained by Sennah & Kennedy (1999) was used to determine the torsional stiffness of the prototype bridges using membrane analysis method were employed in this research:

- For two-box bridges:

$$J = \frac{8z^2 t_4}{X + Y} \qquad (2.33)$$

- For a three-box bridges:

$$J = \frac{4z^2 d' (3X + 10Y)}{X^2 + 4XY + 2Y^2} \qquad (2.34)$$

- For a four-box bridge:

$$J = \frac{8z^2 d' (2X + 5Y)}{X^2 + 3XY + Y^2} \qquad (2.35)$$

- For five-box bridge:

$$J = \frac{4z^2 t_4 (5X^2 + 28XY + 35Y^2)}{(X + 2Y)(X^2 + 4XY + Y^2)} \qquad (2.36)$$

where $X = B [1 + (d'/d'')]$; $Y = d [(d'/t_w)]$; and $Z = Bd$. In the Figure 2.61 d', d" and t_w were identified.

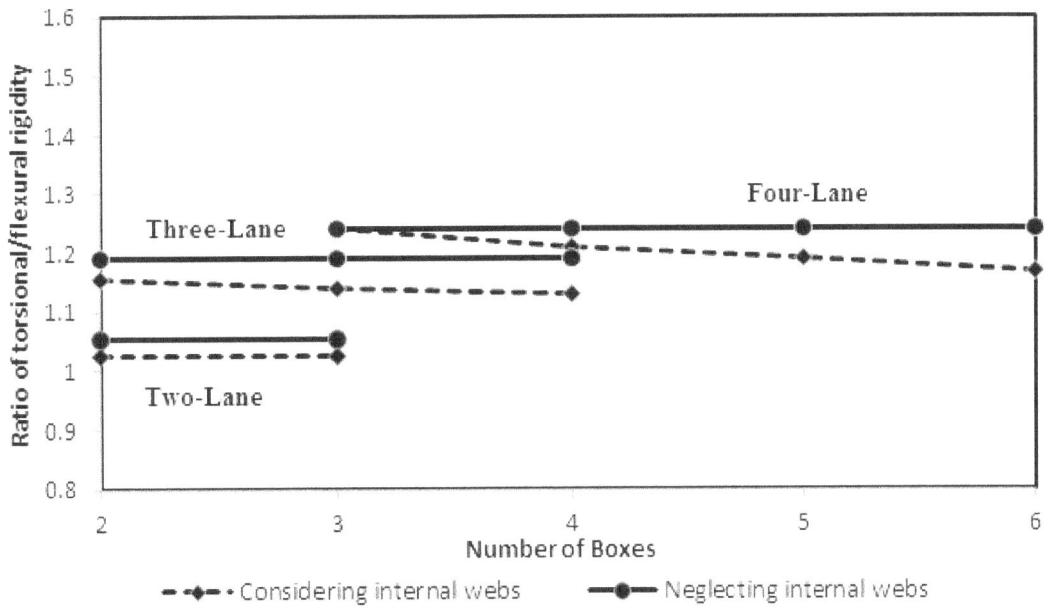

Figure 2.67 Effect of Number of Boxes on Torsional Stiffness of MCB Bridges

2.21.2 Effect of Torsional Stiffness

In bridge design procedures, although the influence of torsional moment on determining design actions is underestimated, the twisting moment induces transverse and negative moment and shear of skewed bridges that must be considered (Théoret et al. 2011)

The effect of torsion on distribution of bending moment and shear in skewed bridges was considered. The preliminary study indicated that changing concrete slab thickness, span-to-depth ratio and top and bottom flange thickness do not have significant effects on skewed bridge actions(Huo et al. 2003; Song et al. 2003). However, it was observed that torsional and flexural rigidity of bridges intensified the effects of the above parameters.

Torsion induces the bending moment and shear in skewed bridges under traffic loading conditions. Subjected to flexural loading, the multicell box section deflects stiffly (longitudinal flexure) and deforms (normal distortion); under twisting loading, the box section rotates rigidly and deforms (warping distortion). If adequate intermediate diaphragms in size and number are provided, the effects of sectional distortion can be disregarded.

Figure 2.68 through Figure 2.71 show the influence of torsional stiffness (J), on the bending moment, shear and twisting moment of skewed bridges. The results were presented as a function of the ratio of the bending moment, shear and torsion in skewed bridges to the corresponding straight bridge values. Four levels of torsional stiffness corresponding to: (1) ultimate limit state ULS, (2) serviceability limit state SLS, (3) an arbitrary value (20% of J), and (4) torsion-less state were applied. In the case of SLS, uncracked section with full stiffness was assumed, whereas in the ULS condition, it was assumed that the bridge's decks were severely cracked and then only 50% of the torsional stiffness was applied (Euro-code 2 2005; AASHTO LRFD 2008).

Figure 6.68 indicates that torsional stiffness has a negligible effect on the positive bending moment (maximum 10%), thus the additional reinforcement on the longitudinal direction to prevent cracks developing is not necessary.

The effects of torsional stiffness on the negative bending moment is shown in Figure 2.69 where the maximum 3.6 for the ratio M_{skew}/M_0 revealed approximately a 32% reduction in negative flexural bending in a case with a 60° skew angle. Great attention was paid to the obtuse corner of the skewed bridges where the maximum negative moment and shear were observed. Therefore, additional reinforcement was provided at this location to prevent crack development.

For skewed bridges, the maximum shear accrues in the external girders at the obtuse corner (Huo & Zhang 2008). In the serviceability limit state condition, the maximum shear was obtained. However, with decreasing torsional stiffness, the maximum shear for lower skew angles trends to occur in first internal girder (Figure 2.70). With an increase of the skewness, the positive bending moment near the mid-span decreases and the negative bending moment near the supports increases.

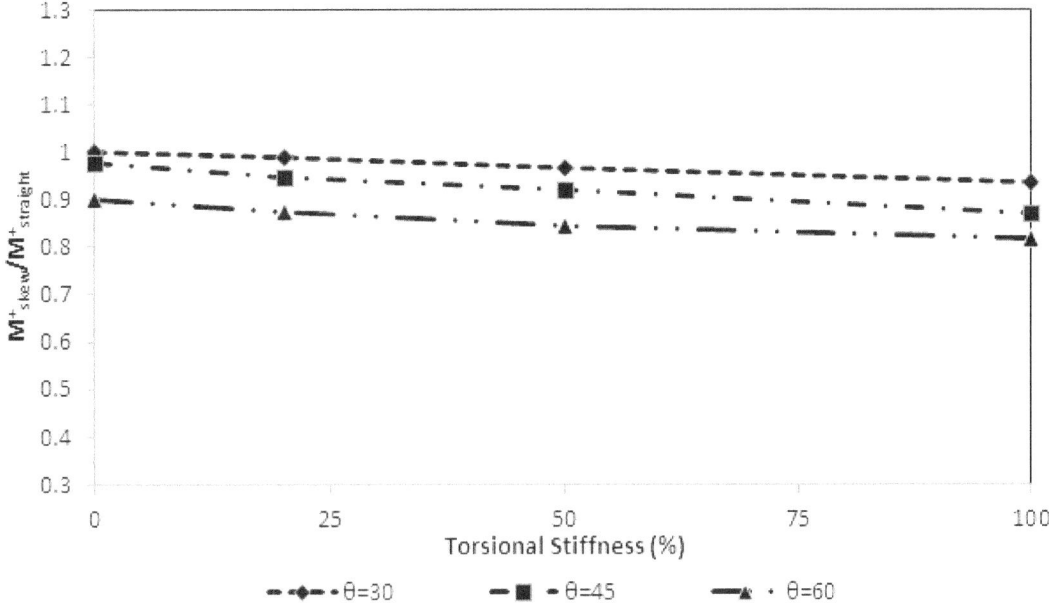

Figure 2.68 Torsional Stiffness vs. Maximum positive Bending Moment

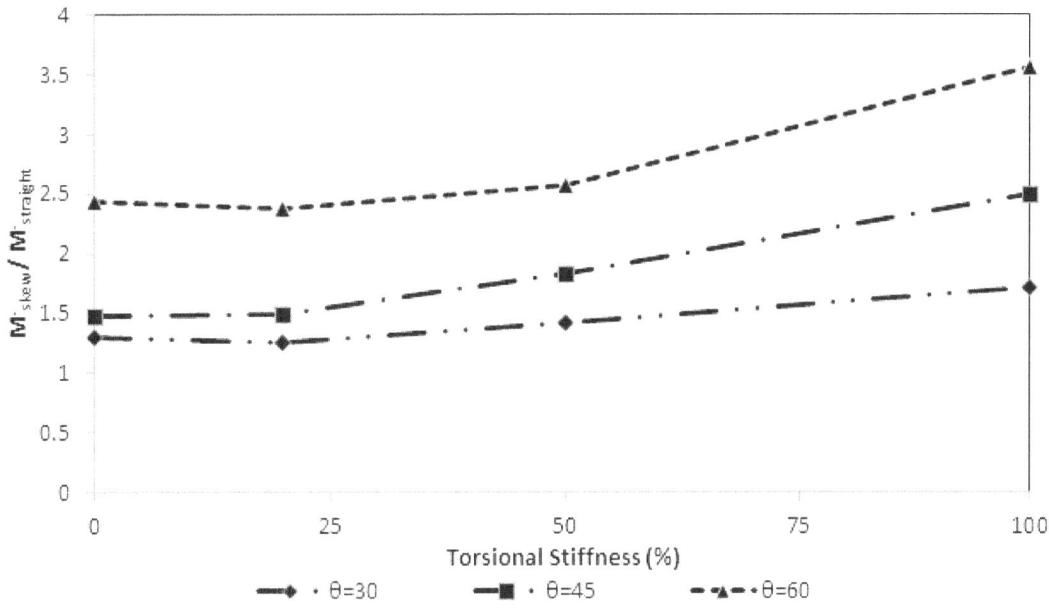

Figure 2.69 Torsional Stiffness vs. negative Bending Moment

In addition, the maximum torsional moment of skewed and straight bridges (S_{skew} and $S_{straight}$) occurred in different girders due to the location of wheel loads on the superstructure, drawing charts as a function of $S_{skew}/S_{straight}$ may lead to ridiculous results. Hence, Figure 2.71 was plotted based on the relationship between torsional stiffness and the positive torsion at the exterior graders for different skew angles.

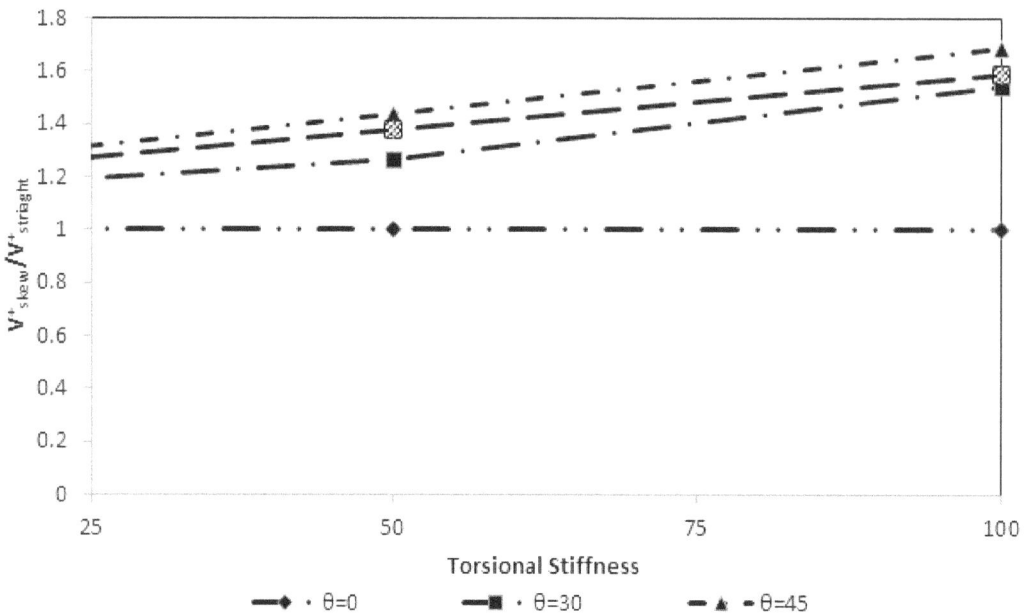

Figure 2.70 Effect of Torsional Stiffness on Distribution of Shear Force

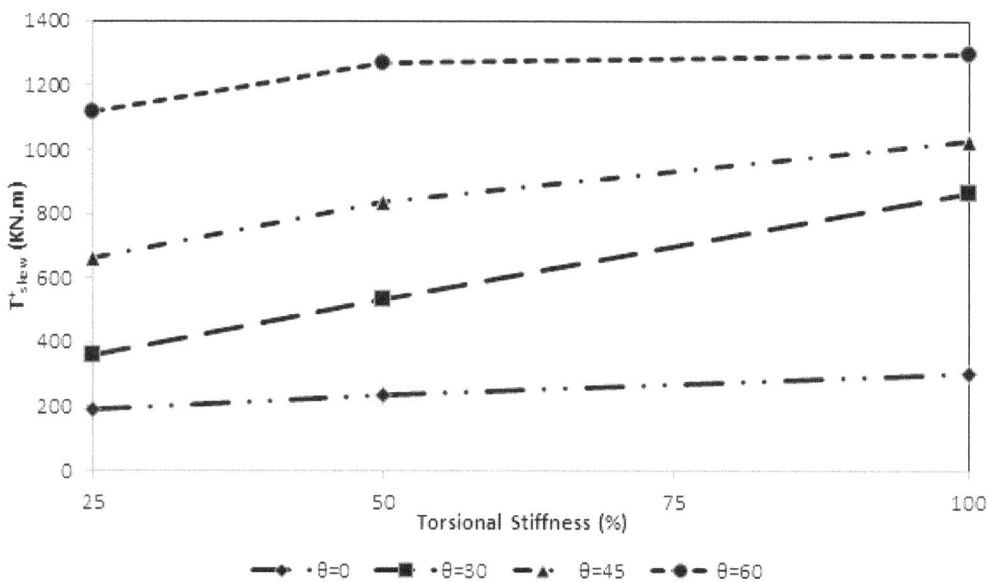

Figure 2.71 Torsional Stiffness vs. Live Load Distribution of Torsion

Figure 2.71 indicates that positive torque decreased by up to 22% for bridge with skew angle of 60°, when torsional stiffness decreased. Diagonal cracking due to exceeding the diagonal principal tension from tensile strength was the main cause of this reduction; nevertheless, after cracking, the post cracking behaviour, such as tension-softening in both reinforced and unreinforced concrete regions was considerable. Unfortunately, the majority of finite element programs are unable to take into account this phenomenon during bridge analysis (Krätzig 1993).

The effect on the bridge response of the presence of the intermediate diaphragms was also evaluated. Figure 2.69 and Figure 2.70 show that the variation in bending moment and shear of prototype multicell box girder bridges with intermediate diaphragms (IDs) significantly improves. Similar improvement was observed for torsion distribution on the superstructure. The same finding has been observed by other studies (Zokaie et al. 1993; Foinquinos et al. 1997; Huo et al. 2005).

2.22 Longitudinal Distribution of Twisting Moment

Figure 2.72 shows the distribution of torsion on the girders of 60 m multicell box-girder bridges with a 30° skew angle. The bridge was loaded with a design truck (HS 20-44) at the mid-span of the first lane of the superstructure.

The location of the truck is shown in Figure 2.72 and the distribution of twisting depends on the distance from the centre of gravity of truck and girders. The maximum positive twisting moment occurred at the obtuse corner of the first intermediate girder at a distance of 2.90 m from the centre of gravity of the live load. Consequently, the middle and third girders gained the lower proportion of total twisting moment, respectively. The maximum twisting moment was obtained at the acute corner of right external girder (at a distance of 0.60 m from the truck), which is significantly higher than the twisting moment of the other girders.

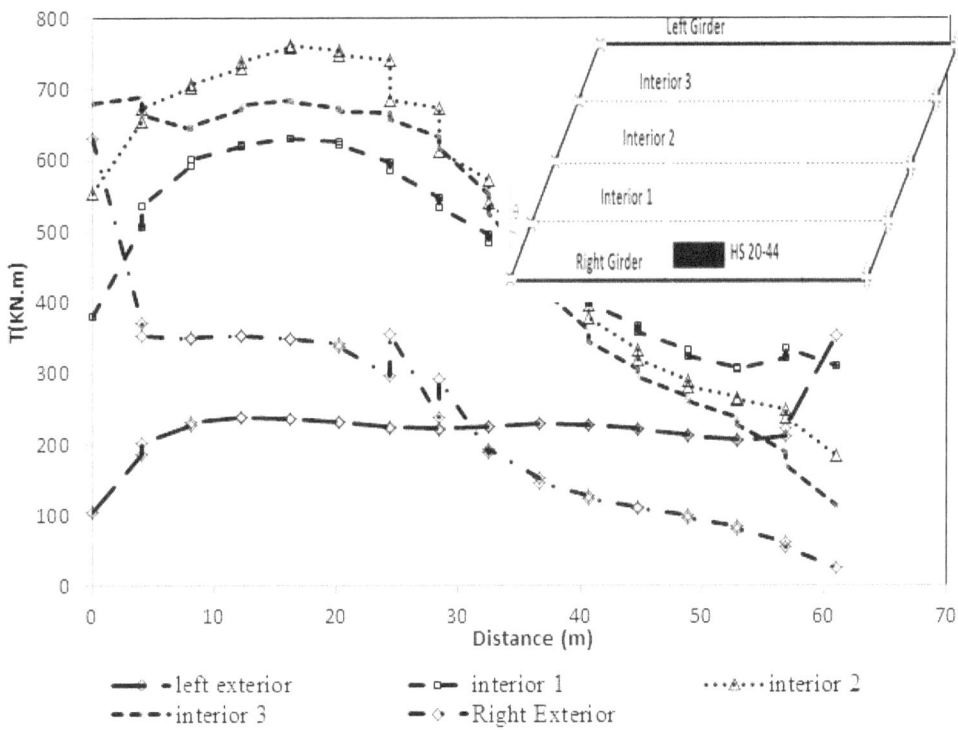

Figure 2.72 Distribution of Twisting Moment in Longitudinal Direction of a Skewed Bridge subjected to Truck Hs-20-44

In addition, when driving the design trucks to the right of their respective lanes, it could be observed that the positive torque at the obtuse corner was reduced and that the negative torque at the acute corner was increased. Similar results were obtained when increasing the number of trucks in the transverse direction. The similar observations were obtained for continuous bridges.

2.23 Summary

The live load distribution factors in continuous straight and skew cast-in-place concrete multicell box-girder bridges under AASHTO LRFD designated live load were investigated. Parametric study was performed to determine the maximum tensile, and compressive stress and deflection employing the adjusted finite element model of bridges.

The finding results from parametric study indicated that the span length of bridge, number of lane loads, number of boxes, and skew angle are the most critical parameters affecting the live load distribution factors.

The proposed equations for aforementioned live load distribution factors were derived for straight and skew two span continuous concrete multicell box-girder bridges. It was indicated that these empirical equations have been sufficiently accurate, easy to use, and obtain details inaccessible in the current bridge design specifications. The proposed equations obviate the necessity for the rigorous analysis including the finite element analysis of bridge for various loading conditions. The proposed equations for live load distribution factor can be employed for design of bridge with two to four spans.

The influence of intermediate diaphragms on live load distribution factor for bending moment was qualified in this chapter. Finding results from initial parametric study indicated that the intermediate diaphragms effect on bridge construction is significantly depends on span length, skew angle, and intermediate diaphragms arrangement. It was concluded that the deck with intermediate diaphragms perpendicular to the longitudinal webs are the best positioning. Using this internal diaphragms arrangement, the influence of number of diaphragms on live load distribution factor is insignificant.

Modification factor expressions for intermediate diaphragms effectiveness for both external and internal girders were deduced based on a statistical analysis on obtained results from finite element analysis. The accuracy of proposed expressions was verified by conducting a comparative study between obtained results from empirical expressions and those obtained from various analytical methods. Based on the finding results from finite element method and proposed equations, it was concluded that intermediate diaphragms decreases the live load distribution factor of bending moment.

To improve the accuracy of shear and moment distribution factors from simplified Henry's method, two set of skew correction factor expression in term of LRFD skew correction factor formulas were deduced. The results of parametric study indicated that the main parameters affecting the skew correction factor of shear and bending moment are including skew angle, span-to-length ratio, and girder spacing. Finding results from parametric study were used to deduce skew correction factor expressions for single span and continuous multicell box girder bridges. The slightly higher than unity average and small standard deviation and coefficient of variation for proposed expressions verified the accuracy and reliability equations.

The distribution of torsional moment on skew bridges was evaluated. It was concluded that torsional stiffness has a significant effect on bridge actions, but the effect of internal webs on torsional stiffness is negligible. The twisting distribution at the obtuse corner is higher than that at the acute corner of skewed bridges and with an increasing skew angle, the twisting moment at the girder increases significantly.

NUMERICAL MODELING AND PARAMETRIC STUDY OF SKEWED MULTICELL BOX-GIRDER BRIDGES

Three-dimensional finite element model was employed to evaluate static and dynamic behaviour of straight and skewed multicell box-girder bridges. A full SAP2000 (CSI 2009) finite element model was used for loading conditions that needed more detailed analysis, such as the bridge-vehicle interaction and bridge deck under heavily truck loading. A simplified finite element modeling method was also used to conduct parametric studies and to evaluate the effectiveness of bridge parameters on live load distribution and dynamic impact factors of selected bridges. The finding results, then, applied to formulate accurate equations for both live load distribution factor and dynamic impact factor of multicell box-girder bridges. In the following, the methodologies implemented to achieve aforementioned objectives were discussed.

The parametric studies were carried out on 280 prototype multicell box-girder bridges to determine the key parameters affecting on static and dynamic distribution of live load on the bridges. The parameters eveluated in this study were; the number of boxes NB, width of bridge deck W, skew angle θ, span length L, the number of lane loaded NL, continuity, and intermediate diaphragms IDs. All these parameters were changed to examine the effectiveness of each parameter on bridge responses. The parameters are appropriately selected in the practical ranges of these variables in other to cover all possible configurations of bridges.

3.1 Geometric Configuration of Bridges

In this study, two sets of prototype multicell box-girder bridges (MCB) were design and modeled. In the first set, width of the bridges is taken as 9.10 m for two lanes, 14.00 m for three lane, and 17.10 m for four lane loaded superstructures as shown in Figure 3.1.

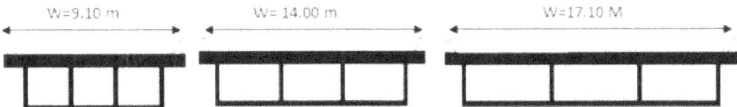

Figure 3.1 Cross Section Configurations with Various Widths for Three-Box Bridges

For placing the vehicle loading near to the edge, 0.5 m thick barrier was supposed along the each edge; however the barriers were not simulated in bridge modeling. The preliminary studies indicated that the slab thickness has an insignificant impact on live load distribution factor of bridges, so in the first sets of bridges, the upper and bottom slab thickness for all prototype bridges were taken as 0.20 m and 0.15 m, respectively. The second set of bridges were developed using Ontario method to developed proposed equations for skew correction factor in term of LRFD formulas. Parameters involved in this study are:

a) Span length of bridge, the maximum and minimum value of the span length was chosen in excess of limitation on the range of applicability defined in the AASHTO LRFD (2008) specifications. When engineers design a bridge with structural properties in exceed of restricted ranges of applicability, the simplified formulas in the AASHTO LRFD (2008) specifications no longer are feasible.

b) The number of boxes of two to six was selected so that supply the practical ranges of actual skewed and straight multicell box-girder bridges, and in exceed of the limitation of AASHTO LRFD (2008) specification on girder spacing.

c) Skew angle, to consider the effect of skewness on dynamic and static behaviour of bridges; the skew angle was varied from 0 to 60°, with 15° increments.

d) Continuous bridge was also considered in this study.

e) All prototype bridges were simulated and analyzed without intermediate diaphragms IDs, similar to that used by Zokaie et al. (1993) to formulate the live load distribution factor of bending moment and shear force adopted by LRFD specifications, and then with solid concrete intermediate diaphragms IDs. The different number of intermediate diaphragms are applied on bridges to consider the most efficient arrangement of this secondary elements, particularly in skew bridges.

f) The effect of torsional stiffness on skewed bridges is considered in this study. To this purpose, four level of torsional stiffness corresponding to; ultimate limit state USL; serviceability limit state SLS; an arbitrary value (20% of J); and torsionless state were applied. In the case of SLS, uncracked stiffness were assumed, however, in the ULS condition, it was assumed that bridge's decks severely cracked, then only 50% of torsional stiffness were applied (Euro-code 2 2005; AASHTO LRFD 2008)

3.2   Loading Conditions

The HL 93 live loading condition of AASHTO LRFD (2008) specifications, consists the HS 20-44 truck plus design lane load or tandem plus lane load, whichever were govern, used to evaluate the live load distribution factor of straight and skewed multicell box-girder bridges. The bridge was loaded with a series of moving loads, including one truck, two trucks, three trucks, and four or more trucks in order to search for the maximum responses. The trucks were placed at various locations in the longitudinal and transverse directions until the maximum response was obtained. In other word, for each bridge at least 10 different truck positions on transverse direction of bridges are considered. It means more than 2800 analysis are performed.

To evaluate the dynamic bridge-vehicle interaction and develop the dynamic impact factor of skewed and straight multicell box-girder bridges, only HS 20-44 standard truck loading was employed (Samaan 2004; AASHTO LRFD 2008). In this effort, each axle of truck was idealized as a pair of concentrated loads passing over the bridge deck on a specified lane with a constant speed. The mass of trucks were neglected in the dynamic analysis because in all cases, the mass of bridge superstructure was more than 10 times than mass of truck (Ashebo et al. 2007; Moghimi & Ronagh 2008).

The minimum spacing of 600 mm was set between the truck wheel and curb, and the minimum distance of wheel line of two laterally adjacent trucks was 1.20 m, as recommended by AASHTO LRFD (2008) specifications. The information in regard to longitudinal and transverse position, and number of applied trucks on superstructure thoroughly describe in Chapter II.

3.3   Computing of Live Load Distribution Factor

The live load distribution factors for the bridge straining actions were obtained by dividing the maximum responses of bridge obtained from refined method of three-dimensional finite element technique, with maximum responses of the corresponding bridge action for a simulated single beam. The simulated beam was drawn up by dividing the multicell box-girder cross section of the entire bridge into a number of distinct I and T shaped beam as indicated in Figure 3.2.

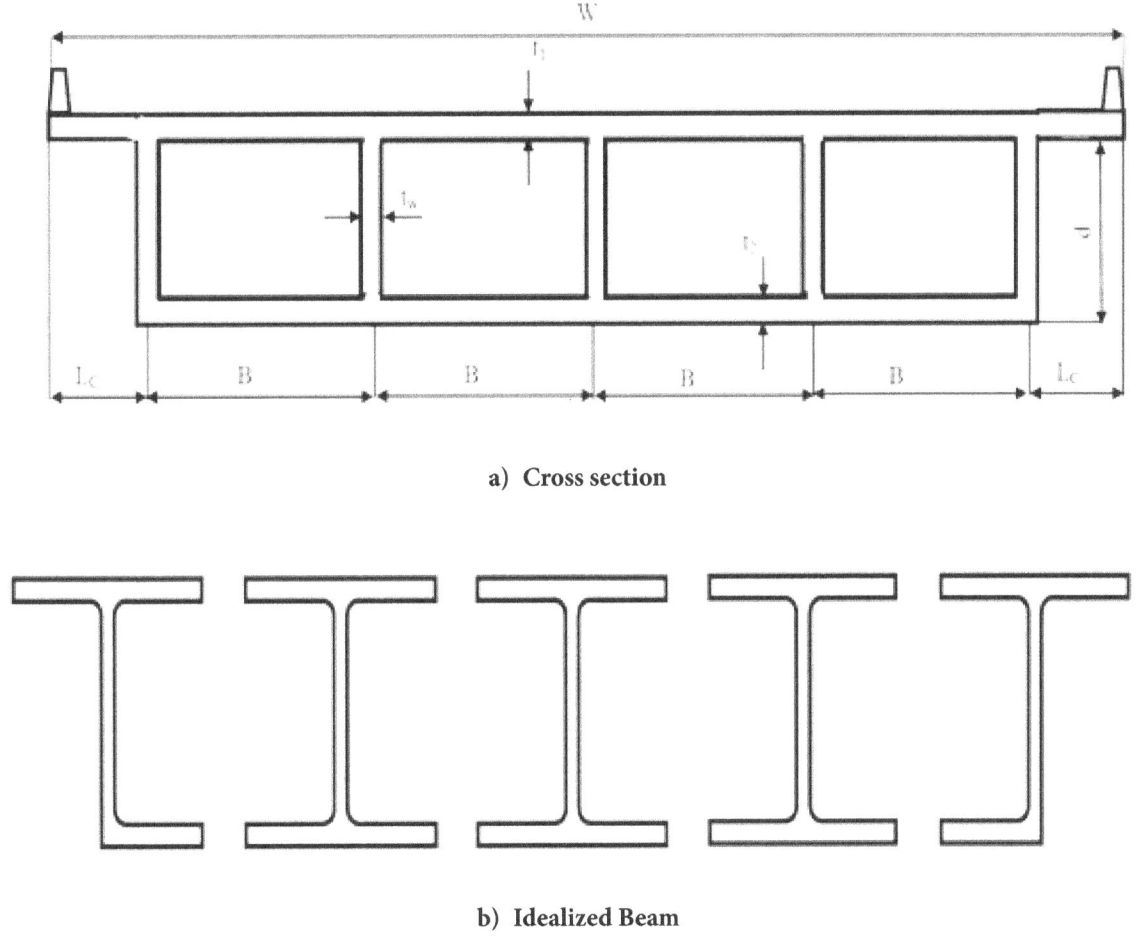

a) Cross section

b) Idealized Beam

Figure 3.2 Cross Section Symbols and Typical Simulated of a Four-Box Bridges

Each simulated beam includes of one concrete web and upper and bottom flanges, W/(1+NB) portion of concrete slab, where W is bridge width and NB stands number of boxes. The maximum straining actions for the simulated beam was determined by using the simple-beam bending theory.

The described method were used in presented study to obtain the maximum distribution factor of compressive stress at the intermediate supports, tensile stress at mid-span, reaction at the piers and deflection at the mid-span of bridges.

The maximum tensile stress σp, was determined near the mid-span using finite element analysis of SAP2000 (CSI 2009) program. Then the maximum stress at the bottom fiber close the mid-span σpa, for simulated single beam was obtained subjected to one line of truck wheel. Then, the maximum tensile stress distribution factor, Dσp, was determined from following expression:

$$D\sigma_p = \frac{\sigma_p}{\sigma_{pa}} \qquad (3.1)$$

In the same manner, the live load distribution factor of compressive stress, $D\sigma_n$ at the intermediate supports of continuous bridge was determined. The maximum compressive stress at upper fiber at the pier, σ_{na}, for the simulated beam was obtained using the simple beam theory from following equation:

$$\sigma_{na} = \frac{P}{A} + \frac{M}{S} \qquad (3.2)$$

Where A = Area of the simulated beam cross-section, and S =section modulus at the bottom of the beam cross section. Thus, the live load distribution factor for compressive stress,

$$D\sigma_n = \frac{\sigma_n}{\sigma_{na}} \qquad (3.3)$$

The maximum live load distribution factor of deflection was also determined in a same manner as that for stress distribution factor, as following expression:

$$D\delta = \frac{\delta}{\delta_a} \qquad (3.4)$$

Where δ_a is the maximum deflection at the mid-span of the simulated beam, and δ is maximum deflection obtained from refine methods using SAP2000 computer program.

3.4 Formulated Live Load Distribution Factor Equation

A statistical analysis were performed on finding results from parametric study to develop empirical equations for estimating distribution of stress, deflection, and reaction at piers on multicell box-girder bridges. The empirical equations were in term of main parameters that influence the live load distribution, or modification factors of bridges.

To determine the impact of each parameter on distribution of live loads, bridge variables were changed one at a time, and live load distribution factor was obtained for all prototype bridges. Variation of distribution factor of each bridge action with each parameter indicated the significance of that variable. By catching the variation of live load distribution factors with each of significant parameters, empirical equations were derived. This approach was already employed by Zokaie et al. (1993) on NCHRP 12-26 project is expressed as following:

First, it supposed that influence of each parameters could be simulated by functions $f(x_1)$, $f(x_2)$, $f(x_3)$, and $f(\ldots)$, respectively. The live load distribution factor, LDF, stated in the following expression:

$$LDF = a. \ f(x_1) f(x_2) f(x_3) f(\ldots) \qquad (3.5)$$

where 'a' is the scale factor to be obtained on the basis of the variations of live load distribution factors with these parameters.

Regression analysis was used to obtain the best expression to be compatible with the variation of live load distribution factor, and each independent parameter. The "coefficient of determination R^2" was practiced in the context of statistical models to evaluate how well outcome were likely to be estimated by the model. For each variable, linear, exponential, and polynomial functions were examined in order to find the one with best coefficient of determination (nearest to unity). In some cases, when the preciseness is satisfactory, the uncomplicated expression was adopted.

This method widely used by current bridge design codes and researcher in order to deduced new practical formulas for bridge design. For example, the AASHTO LRFD (2008) specification and Zhang (2008) used the following function to develop the live load distribution factor and skew correction factor of bridges:

$$LDF = (a)(x_1^{c1})(x_2^{c2})(x_3^{c3})(....) \qquad (3.6)$$

where x_1, x_2, x_3 are the variables comprised in the formulas, 'a' is scale factor and c1, c2, c3 are obtained in term of x_1, x_2, and x_3, respectively. If variable x_1 changes from x_{11} to x_{12}, the live load distribution factor changes from LDF_1 to LDF_2, while other parameters retain constant.

$$LDF_1 = (a)(x_{11}^{c1})(x_2^{c2})(x_3^{c3})(....) \qquad (3.7)$$

$$LDF_2 = (a)(x_{12}^{c1})(x_2^{c2})(x_3^{c3})(....) \qquad (3.8)$$

Thus

$$\frac{LDF_1}{LDF_2} = (\frac{x_{11}}{x_{12}})^{c1} \qquad (3.9)$$

$$c1 = \frac{Ln\frac{LDF_1}{LDF_2}}{Ln\frac{x_{11}}{x_{12}}} \qquad (3.10)$$

If n different values of 'x_{1i}' are obtained and sequential pair are utilized to calculate the amount of c1, therefore, (n-1) distinctive values for this variable would be achieve. The average of (n-1) values of c1 is employed to attain the best coincidence.

In some cases, an exponential expression is not match to simulate the influence of a variable. Thus other functions should be considered. As an example, for one lane loading multicell box-girder bridges, the best function to express the live load distribution factor of bending moment is as following (Zokaie 2000):

$$LDF = (a)[x_1 + c1](x_2^{c2})(x_3^{c3})(....) \qquad (3.11)$$

where c1 is a constant to be acquired established upon the variation of live load distribution factor with variable x_1. Supposing that all bridge variables, except x_1, retained at the average amount, then:

$$LDF = (a)(x_{11} + c1)(x_2^{c2})(x_3^{c3})(...) \qquad (3.12)$$

$$LDF = (a)(x_{12} + c1)(x_2^{c2})(x_3^{c3})(\ldots) \qquad (3.13)$$

Thus

$$\frac{LDF_1}{LDF_2} = \left[\frac{c1 + x_{11}}{c1 + x_{12}}\right] \qquad (3.14)$$

$$c1 = \frac{L_2 x_1 - L_1 x_2}{L_1 - L_2} \qquad (3.15)$$

The value of c1 can be determined in the same manner of Eq. (3.10).

3.5 Dynamic Impact Factor

One of the main objectives of this book is the development dynamic impact factor (DIF) or dynamic load allowance (DLA) of multicell box-girder bridges. To achieve the most crucial vehicle loading conditions and truck position in the lateral direction, a skewed and a straight multicell box-girder bridge with various truck positions were examined. The maximum safe allowance highway velocity of 120 km/h which adopted in most specifications was used in this research.

The permissible truck speed can be increased with raising the amount of friction. Thus the highest coefficient of fraction of 0.20 is applied for all simulated bridges in this boo. It is because the friction between the tires of truck and bridge surface which widely depends upon the quality of asphalt, bridge surface condition, vehicle speed and tire, has a significant effect of motorist comfort.

A fully loaded skewed and a straight multicell box-girder bridge were analyzed using mode superposition method and direct integration technique to determine the most sufficient preciseness and rational computation time in the dynamic analysis of box-girder bridges. The selected method was used to conduct a sensitive parametric study to evaluate the effect of main parameters on dynamic bridge-vehicle interaction. The vehicle speed, number of boxes, skew angle, number of boxes, and number of lanes were investigated.

The finding results of the parametric study were used to develop the empirical expressions for dynamic impact factor DIF, or dynamic load allowance of multicell box-girder bridges. The AASHTO (2002) standard specifications state that the dynamic impact factor obtains from following expression:

Maximum dynamic response = maximum static response (1+DIF)

While the dynamic impact factor is assigned as:

$$DIF(\%) = \frac{R_{dynamic} - R_{static}}{R_{static}} \times 100 \qquad (3.16)$$

where $R_{dynamic}$ and R_{static} are the maximum dynamic and static bridge straining actions obtained from refine methods. The maximum dynamic and static bridge responses for tensile and compressive stress, deflection, reaction at piers, shear force, and bending moment were calculate using Eq. (3.16), for the same loading conditions for both analysis.

As mentioned earlier, one of the main objectives of this study was to establish simplified empirical expressions for dynamic impact factor of bridges. Thus, the variations of dynamic impact factor from finite element analysis using SAP2000 programs, versus first fundamental and span length of bridges were plotted. Upper bound values of dynamic impact factor were used to obtain two sets of empirical expression for each bridge actions including, stress, bending moment, torsion, and frequencies.

3.6   Finite Element Modeling of SAP2000 Software

SAP2000 is a user friendly engineering modeling software established upon three-dimensional finite element technique that easily resolve elastic and non-elastic problems regarding bridges.

The software includes an extensive collection of elements that make it capable to simulate almost all arbitrary geometric and boundary conditions of bridges. The software also contains a wide range of material properties that can model the behaviour of almost any engineering materials, such as reinforced concrete and steel. The different type of load, such as live load (vehicle, earthquake and wind) and dead load can be simulated in modeling process using "Load Case" option.

A privilege of SAP2000 software, compared with other three-dimensional finite element programs such as ANSYS and ABAQUS, is that it can simulate vehicle loads on defined lane on bridge's deck automatically. The maximum bridge bridges responses such as bending moment, support reactions, shear forces and torsional moment in both longitudinal and transverse directions can be defined without simulating the truck loads manually.

The three-dimensional model was applied to simulate skewed and straight multicell box-girder bridges in SAP2000 program. A four node shell element with six degree of freedom at each node was employed to model the superstructure of prototype bridges. Upper and bottom elements of web were integrated with the top and lower slabs at connection points to ensure compatibility of deformation. In nature, with increase the number of shell elements for a model, the processing time general. For this study, the dimensions of shell varied based on the skew angle. For straight bridge, the shell elements with size 4 m×4 m were applied. Due to complexity of skew bridges, the smaller dimensions ware selected for meshing the models with shell elements. Based on the NCHRP 12-26 (Huo et al. 2003) skew bridges, the shell elements with dimensions 3m×3m were used.

The transvers solid end diaphragms at piers and abutments were simulated using the shell elements with the size and properties of designated diaphragms (Huo & Zhang 2007).

3.6.1   Material Properties

The material library in SAP2000 contains a large range of materials that enable the simulating of the material properties used in the bridge. In addition, the program requires that properties sufficiently define to obtain accurate results from bridge analysis. The materials associated with the bridge element should contain a mechanical category option to explain its mechanical (stress-strain) properties such as elasticity modulus and Poisson's ratio.

3.6.2   Boundary Conditions

The multicell box-girder bridges in the presented investigation was simulated such that the longitudinal axis was along the global X direction, the transverse axis was along the global Y direction, and the vertical axis was along the global Z direction. The support at one abutment was assumed to be hinge, and roller supports were assumed for the other piers. At the hinged abutment, the girders of the bridge were constrained against translation in the global X, Y and Z directions. At the roller supports, the girder was allowed from translation in only the X directions. The rotation of supports around the X directions (θ_x) was restrained by providing sufficient solid concrete support diaphragms. Figure 3.3 indicates the accepted boundary condition of skewed and straight multicell box girder bridges.

The adopted finite element modeling method was validated using a comparative study on results obtained from field test study and analytical methods, on skewed and straight multicell box-girder bridges subjected to vehicle loading conditions.

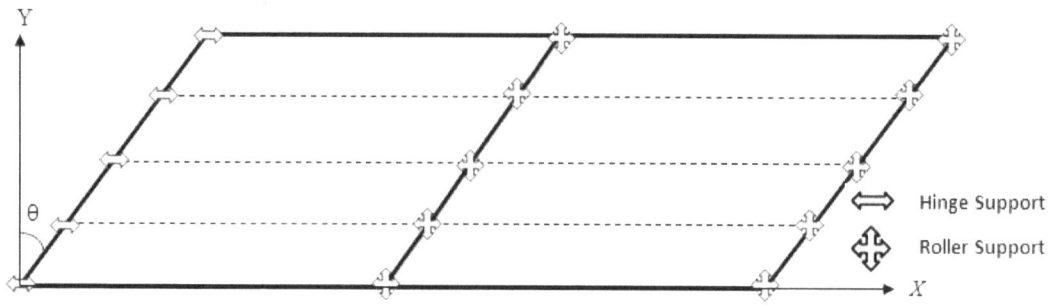

Figure 3.3 Boundary Conditions of a Skew Multicell Box-Girder Bridges

3.7   Analysis Methods of SAP2000 Program

Different methods are available in SAP2000 software to analyses the bridge subjected to static and dynamic loads. The following methods were described;

3.7.1   Static Analysis

The SAP2000 program uses the static procedure "multistep-static" option to conduct the elastic static analysis of bridges. The bridge analysis can be employed to compute influence lines for traffic on bridge superstructure and to analyze these structures for the responses resulting from vehicle loads.

3.8   Dynamic Analysis

A dynamic model of structure is one that contains the influence of damping and mass. The SAP2000 program suggests various techniques for evaluation the dynamic behaviour of bridges. To solve the elastic problems, modal superposition technique or direct integration method may be employed.

3.8.1   Extraction of Fundamental Frequency and Mode Shapes

SAP2000 computer software provides modal analysis to conduct a fundamental frequency extraction. The modal analysis applied the Eigen vectors to obtain the frequencies and mode shapes of selected bridges. Eigen vector analysis determines the undamped free-vibration mode shapes and frequencies of the system. These natural modes provide an excellent insight into the behaviour of the structure. They can also be used as the basis for response-spectrum or time-history analyses, although Ritz vectors are recommended for this purpose.

3.8.2   Transient Modal Dynamic Analysis

A modal dynamic procedure is available in SAP2000 computer program by using dynamic Modal in Time History Option, to examine transient elastic dynamic problems. The modal dynamic procedure should only be fulfilled after obtaining the natural frequency of structure, so that it estimates the dynamic responses based on the defined natural frequencies of the structures. However, this is a well-liked method of dynamic analysis, it has some restrictions. For instance, the method can be used for elastic systems, with proportional damping.

3.8.3   Direct Integration Method

The dynamic behaviour of most elastic and nonlinear systems undergoing time-dependent changes (transients) can be analyzed by applying direct integration method. In this method, the differential equations are combined with the equations of motion, and the deflection is determined directly by applying a numerical step-by-step process. The numerical integration methods widely employed for the approximate solutions by applying the implicit or explicit schemes.

The SAP2000 software use Hiber-Hughes-Taylor operator (Hilber & Hughes 1978) for solutions of both single and multi-degree of freedom systems with numerical implicit integration scheme. This method has two basic characteristics. First, they do not satisfy the differential equations at all-time t, but only at discrete time interval, say Δt apart. Secondly, within each time interval Δt, a specified type of variation of the displacement X, velocity $\dot{X}$, and acceleration $\ddot{X}$ are assumed. In the implicit scheme, the matrix of integration operator should be inverted and a set of simultaneous dynamic equilibrium equations should be solved at each time increment.

The SAP2000 software is capable to establish the direct time increment automatically or manually. The effect of the artificial damping can be specified by period or frequency. The value of this indictor changes from 0, which describes no artificial damping, to −0.33, for system with maximum artificial damping. At the bridge with highest level of artificial damping, the damping ratio of about 6% is obtained, when time increment is 40% of the period of oscillation. Since, the effect of damping cannot be very substantial for real bridges.

3.9   Model Verification

To verify that the adopted three-dimensional finite element model is capable of precisely predicting the static and dynamic responses of selected multicell box-girder bridges, the collected data from previous numerical and experimental studies conducted by Huo et al. (2003), Li (1992) and Ashebo et al. (2007) were used. After the validation, the adopted modeling method was employed in this study to evaluate the live load distribution and impact factor of multicell box-girder bridges.

3.9.1   Finite Element Model of Bridge using ANSYS and SAP2000

Four concrete multicell box-girder bridges investigated by Huo et al. (2003) on project TNSPR-PER1218 were selected for this comparative study. Figure 3.4 shows the typical cross section of bridges. Table 3.1 presents the geometric and cross-sectional properties of bridges.

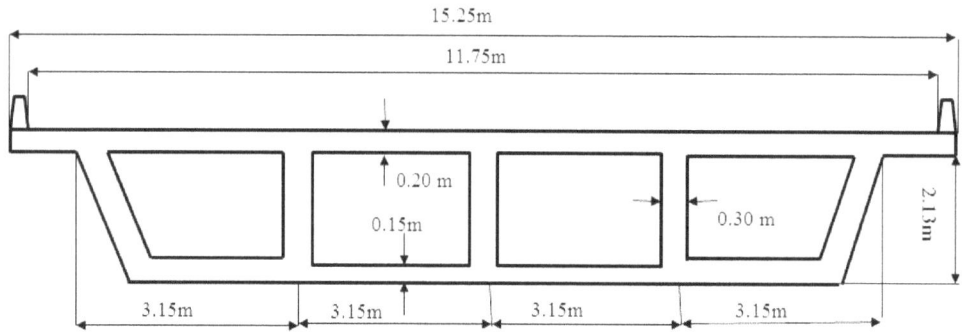

Figure 3.4 Typical Cross Section of Bridge #14 (Huo et al. 2003)

Table 3.1 Dimensions of Selected Multicell Box-Girder Bridge (Huo et al, 2003)

Bridge No.	Span Length (m)			N_B	Skew Angle	Bridge width	Top slab	Bottom Slab	Overhang (m)
	L1*	L2*	L3*						
12	39.00	40.50	-	4	0.00	13.40	0.20	0.15	1.10
13	30.00	30.00	-	4	0.00	13.40	0.21	0.18	1.22
14	27.73	36.30	140.00	4	27.00	15.24	0.23	0.18	1.32
15	33.50	33.50	-	3	16.55	11.00	0.20	0.15	1.14

(* L1, L2 and L3 are the span length of first, second and third spans of each bridge)

Huo et al. (2003) obtained the live load distribution factor of bending moment and shear force for each selected bridges using ANSYS program. Four node SHELL63 elements were used to simulate the concrete deck slabs, webs, and diaphragms. Material properties were set to be linear, elastic, and isotropic. the single beam model (two-dimensional model) and entire bridge model (three dimensional model) were created and analyzed to determine the maximum bridge responses of highway bridges under consideration due to HL-93 truck loading in the AASHTO LRFD specifications.

For three-dimensional models, the bridge was loaded with one, two, and three HL-93 trucks and the location of maximum bridge actions was identified by moving the truck independently as well as together. Figure 3.5 and 3.6 indicate the sample loading configuration for maximum bending moment and shear on straight and skewed bridges, respectively (Huo et al. 2003).

The live load distribution factor was determined by dividing the maximum bridge responses from the three-dimensional models by those obtained from the single beam model. The results of this project were used to improve the accuracy of simplified Henry's method in determining the live load distribution factor of bridges.

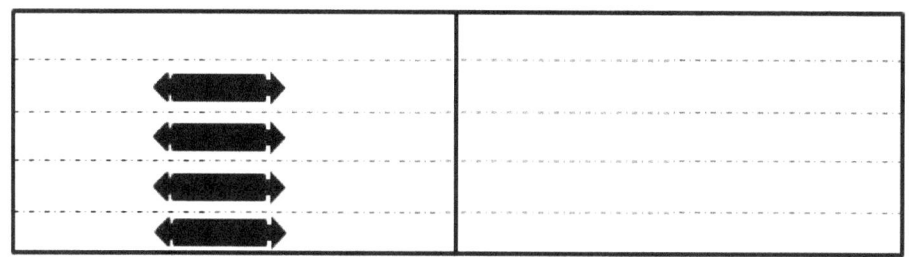

a) Straight bridge

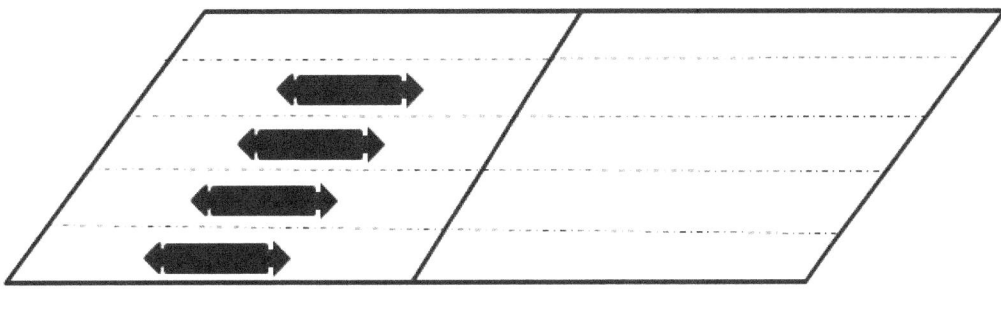

b) Skewed bridges

Figure 3.5 Sample Loading Pattern for Live Load Bending Moment (Huo et al. 2003)

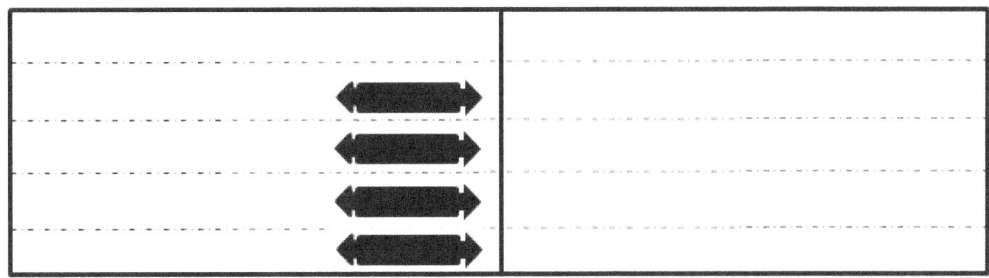

a) Straight bridge

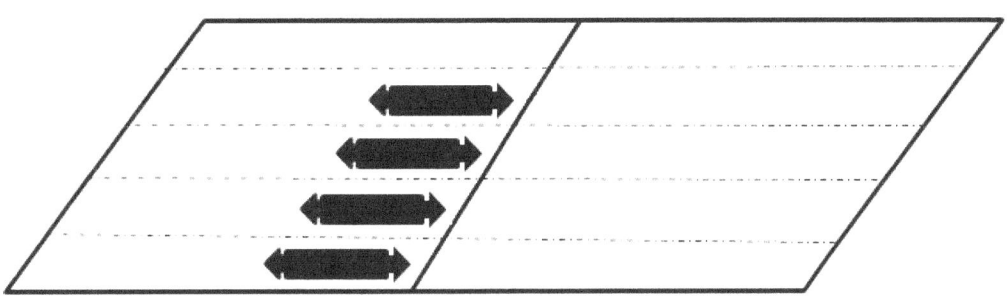

b) Skewed bridges

Figure 3.6 Sample Loading Pattern for Live Load Shear Force (Huo et al. 2003)

The comparison of live load distribution factor (LDF) for bending moment and shear force obtained from finite element modeling of ANSYS and SAP2000 programs are shown to Table 3.2. The errors were calculated on both numerical modeling methods to validate the adopted modeling employing in the finite element analysis. As shown in Table 3.2, the maximum errors for the live load distribution factor from SAP2000 program compared to ANSYS were 7.40% and 4.20% for bending moment and shear force, respectively.

Table 3.2 LDF from ANSYS (Huo et al. 2003) and SAP2000

Bridge No.	LDF of Bending Moment			LDF of Shear Force		
	ANSYS*	SAP2000	Error%	ANSYS*	SAP2000	Error %
12	0.687	0.685	-0.30	0.842	0.852	+1.17
13	0.622	0.672	+7.44	0.856	0.848	-0.94

Contd...

Bridge No.	LDF of Bending Moment			LDF of Shear Force		
	ANSYS*	SAP2000	Error%	ANSYS*	SAP2000	Error %
14	0.665	0.670	+0.74	0.975	1.010	+4.20
15	0.765	0.808	+5.32	0.883	0.904	+2.32

* Source: Huo et al. (2003)

3.9.2 Multicell Box-Girder Bridge Subjected to Concentrated Load

Another objectives of this book is to determine the maximum live load distribution factor of tensile stress at mid-span and compressive stress at the piers of bridges. Thus, the finding results from a study on a two-span two-cell box-girder bridge subjected to self-weight and concentrated loads at mid-span by Li (1992), were used to verify the accuracy of bridge modeling method to estimate the stress and deflection of bridges. The elevation plan and cross-section of bridge is shown in Figure 3.7. The end diaphragms were install at abutments and piers, and loaded eccentrically over the outer webs. The half of bridge superstructure was meshed using 160 thin-walled box beam elements.

The comparison between variations of deflection and distribution of normal stress at mid-span and mid-support of selected bridges were indicated in Figure 3.8 and Figure 3.9. The finding results from plane frame method using SBOXEF program, which developed by Li (1992), and those obtained from SAP2000 computer software, and conventional beam theory were used for this comparative study.

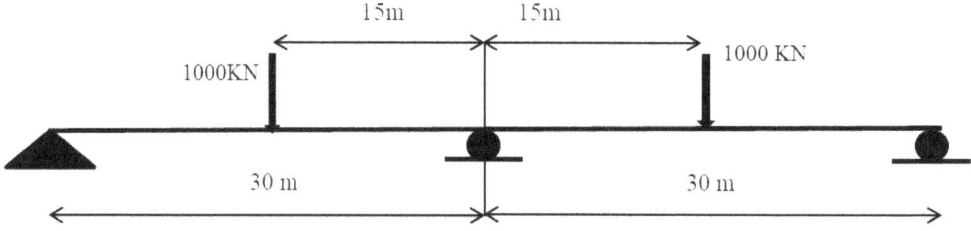

a) **Presentation of Bridge Structure**

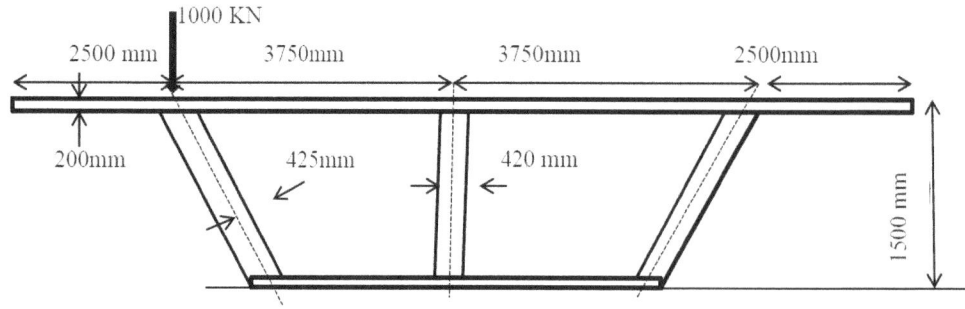

b) **Cross-Section of Bridge Girders**

Figure 3.7 Elevation Plan and Cross-Section of Investigated Bridge (Li 1992)

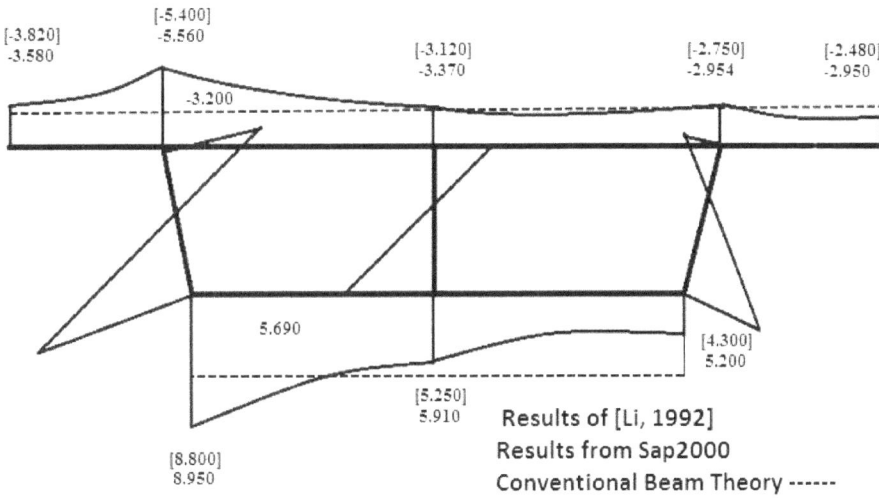

Figure 3.8 Distribution of Normal Stress at the Mid-Support (Li 1992)

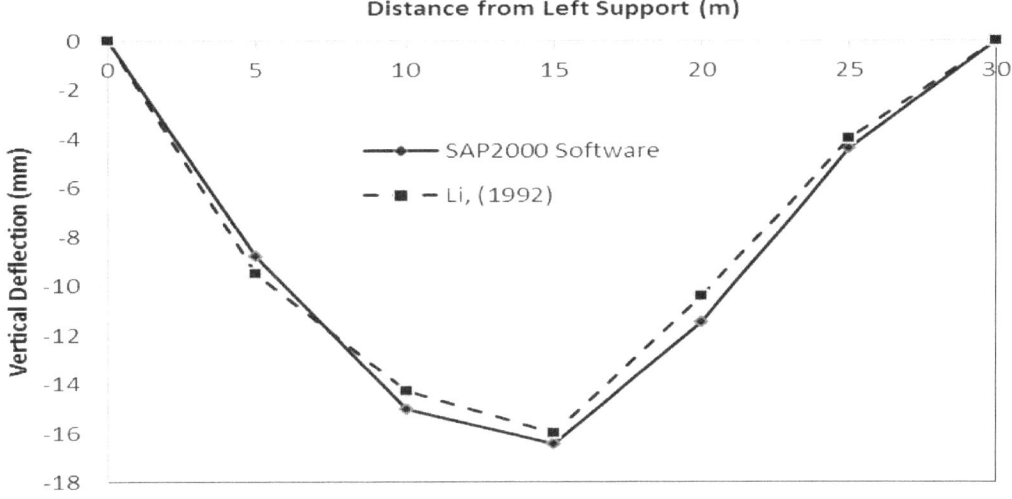

Figure 3.9 Comparison of deflections at the Span of Selected Bridge

From Figure 3.9, it can be observed that deflection and normal stress estimated by previous efforts were relatively close with those determined by adopted herein modeling method. The difference between these two methods is less than 3%. However, the engineering beam theory results diverge from finite element analysis results by up to 30%. It is because the conventional beam theory assumes that distribution of normal stress on cross section of bridge is uniformly.

3.10 Model of Dynamic Analysis of Bridges

Although many experimental studies have been performed on slab-on-girder bridges to evaluate the static and dynamic behaviour, only a few investigations have considered the dynamic responses of box-girder bridges (Ashebo et al. 2007a; Hanna 2008).

The results of a field test study conducted by Ashebo et al (2007a) were employed to verify the results of finite element modeling technique using SAP2000 software. The duplicated Tsing Yi South Bridge in the New

Territories west in Hong-Kong was selected for this study. The bridge is a three-span continues structure with skew angle of 27°, total length of 73 meter, and two lanes with a carriageway. The modulus of elasticity and weight of concrete are 26 Gpa and 24.5 Gpa, respectively. The information about the bridge configuration and boundary condition are presented in Figure 3.10.

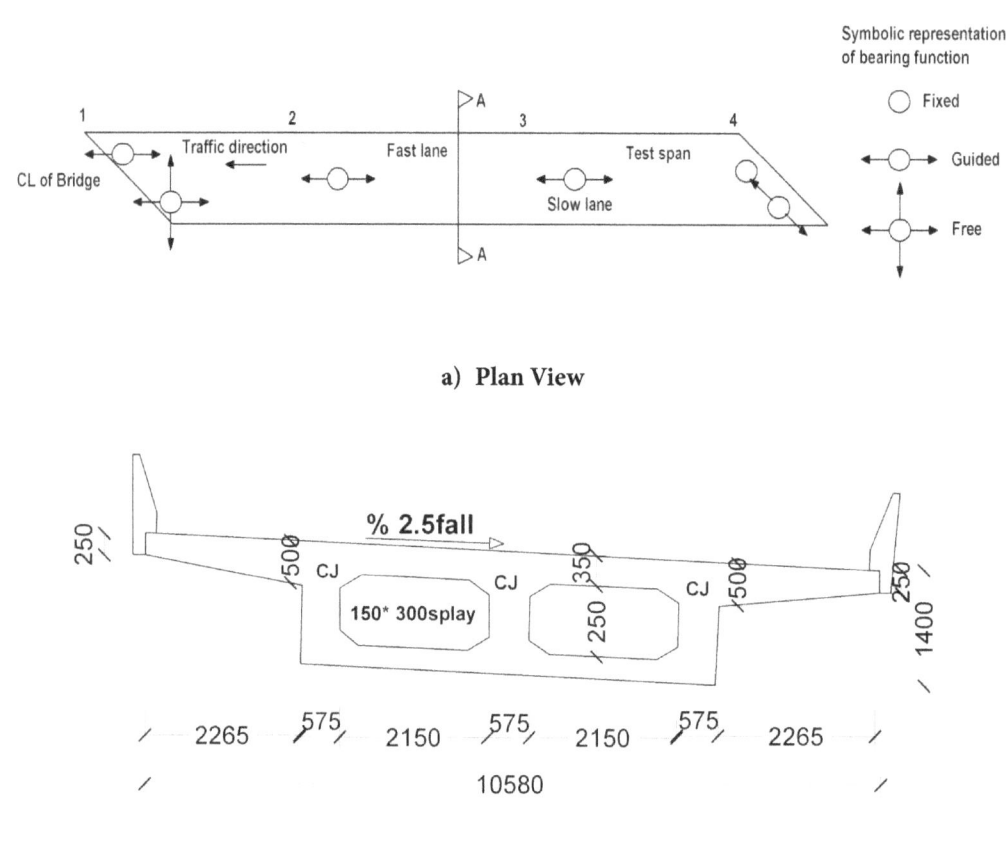

a) **Plan View**

b) **Cross Cross section**

Figure 3.10 Configuration of Tested Bridge (Ashebo et al. 2007a)

A three-axle truck (UD Nissan) with the body width of 2.25 m was selected for field test. The vehicle weighed 24,600 kg, with the weight of the first, second and third axle being 6090 kg, 9410 kg, and 9100 kg, respectively. The spacing between the first and second axles is 4.43 m, and 1.32 m between the second and third axles. The modal test was mainly conducted to obtain the dynamic responses of the bridge such as the fundamental frequencies, and the mode shapes. The technique used in this research was the ambient vibration test. Meanwhile, the dynamic responses of the bridge subjected to controlled traffic conditions were studied. Three modal tests on the bridge were performed at different times including; (1) during 6:30–7:30 a.m which was less traffic flowing on the bridge, (2) during 10:00–10:30 a.m which bridge was under a normal traffic condition, and (3) during 5:15–5:45 p.m. which bridge was under the heaviest flowing of traffic conditions.

The responses of the bridge were measured through six accelerometers (Type ASQ-1LB) with a higher-quality property of lower frequency, which were installed on the soffit of the bridge. The results obtained from the experimental modal test of the three tests and the results obtained from randomly selected controlled traffic conditions are tabulated in Table 4.3. The table also presents the results of modal analysis form three-dimensional finite element modeling method of selected bridge using shell and solid element developed by Ashebo et al. (2007a, b). To verify the adopted herein bridge modeling method, the fundamental frequencies of selected bridge obtained from SAP2000 software were presented in Table 3.3. The bridge was meshed with 396 four node shell elements. It was observed that the finding results from SAP2000 were sufficiently good agreement with field test results so that for most cases the adopted herein modeling method obtain more compatible results than analytical method by Ashebo et al. (2007a, b).

Table 3.3 Frequencies of the Selected Bridge (in Hz) from the field test and FEA

Mode Number			1	2	3	4	5	6	7	8
Ashebo et al. (2007a)	FEA		4.54	4.81	6.70	7.53	10.40	10.68	14.72	16.34
	Field test	1st	4.46	*	6.25	7.82	*	10.87	13.67	15.74
		2nd	4.61	*	6.22	7.73	*	10.81	13.30	15.74
		3rd	4.58	*	6.39	7.76	*	11.11	13.31	15.73
Control Traffic			4.58	*	6.15	*	*	*	*	*
SAP2000			4.28	4.95	5.99	8.71	10.20	11.23	13.43	16.07

(FEA: Finite Element Analysis)

Note: * means undetermined values in field testing

Comparison of the experimental and numerical mode shape and first fundamental frequency are shown in Figure 3.11. It can be seen that the first mode shape from FEA by Ashebo et al. (2007a) was almost 6.5% higher than that from SAP2000 modeling methods. The dynamic impact factor for selected bridge are DIF determined using Eq. (3.16). The dynamic impact factor from field test and finite element analysis were 1.24 and 1.26 respectively, which indicate the adopted finite element modeling agreed well with the results obtained from experimental study.

Generally, the small difference between adopted numerical method and experimental finding indicated that the finite element modeling technique of SAP2000 computer program can reliably obtain the response of multicell box-girder bridges.

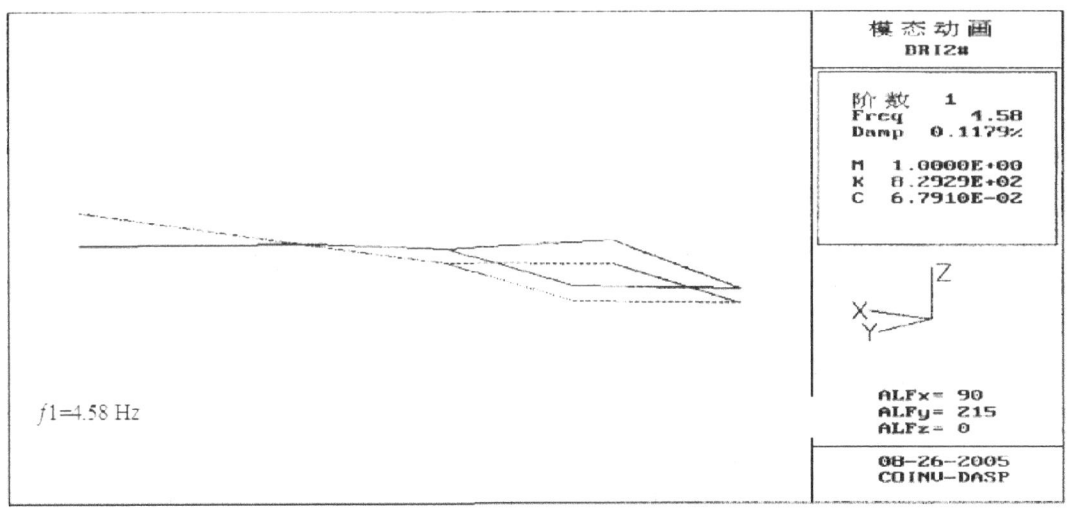

$f1=4.58$ Hz

a) Experimental Study (Ashebo et al. 2007 a)

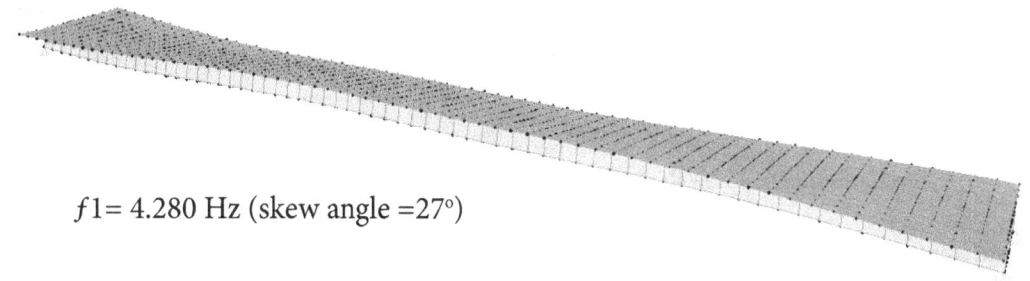

$f1$= 4.280 Hz (skew angle =27°)

b) Numerical Modeling using SAP2000 software

Figure 3.11 First Mode Shapes from the Experiment and Analytical Methods

The step-by-step static analysis was defined to determine the maximum bridge actions subjected to live loads. The modal analysis was employed to extract the fundamental frequencies and mode shapes of bridges and the direct integration method and mode superposition technique available in SAP2000 software, were depicted for evaluating the dynamic behaviour of bridge subjected to moving loads. The simulation of concentrated load, boundary condition and arrangement of intermediate diaphragms were described in the related section.

3.11 Parametric Studies

Parametric studies were conducted on straight and skewed cast-in-place concrete continuous multicell box-girder bridges. The objectives of the parametric studies are: (1) to evaluate the effectiveness of main parameters affecting the maximum bridge actions; (2) to obtain a database for the live load distribution factor, skew modification factor and dynamic impact factor of bridges; and (3) to derive empirical equations for live load distribution factor and dynamic impact factor of multicell box-girder bridges. The HL-93 live loads designated by AASHTO LRFD (2008) were employed in this study.

3.11.1 Bridge Description

In the parametric study for live load distribution factors, the influence of following variables have been considered: (1) Span length, L; (2) Number of boxes, N_B; (3) Number of lanes, N_L; (4) Skew angle, θ; and (5) Intermediate diaphragms, ID_s.

The databases of bridge and bridge design process were developed according to Heins (1978) method and satisfied all range of applicability of AASHTO LRFD (2008) specification. Throughout the study, the length of each span changed from 30 m, 45 m, 60 m, and 75 m to 90m. To evaluate the effect of skewness on structural responses, the skew angle was ranged from 0 to 60° in 15° increment.

The section properties of all prototype bridges employed in this investigation are presented in Table 3.4. In this table: L stands for lane; B stands for box; and the middle number of the nomination indicates the span length of the bridge in meter. For instance, 2L-60-3B means continues two-equal-span bridge of two lanes, three boxes, and each span length being 60 m. The cross-section symbols employ in this study are presented in Figure 3.12, where t1, t2 and t_w stand for thickness of upper and bottom slabs and web, W is the width of bridge, and B stand for web spacing.

Table 3.4 Geometry of Prototype Bridges Used in Parametric Study

Bridge ID	Span Length (m)	No. of Box	No of Lane	Bridge Width (m)	Web Spacing (m)	Bridge Depth (m)	Overhang Width (m)
2L-30-2B	**30.00**	**2**	2	9.15	3.80	1.25	0.76
2L-30-3B	30.00	3	2	9.15	2.50	1.25	0.76
2L-45-2B	45.00	2	2	9.15	3.80	1.90	0.76
2L-45-3B	45.00	3	2	9.15	2.50	1.90	0.76
2L-60-2B	60.00	2	2	9.15	3.80	2.50	0.76
2L-60-3B	60.00	3	2	9.15	2.50	2.50	0.76
2L-75-2B	75.00	2	2	9.15	3.80	3.10	0.76
2L-75-3B	75.00	3	2	9.15	2.50	3.10	0.76
2L-90-2B	90.00	2	2	9.15	3.80	3.70	0.76
2L-90-3B	90.00	3	2	9.15	2.50	3.70	0.76
3L-30-2B	30.00	2	3	14.00	5.80	1.25	1.20
3L-30-3B	30.00	3	3	14.00	3.90	1.25	1.20
3L-30-4B	30.00	4	3	14.00	2.90	1.25	1.20
3L-45-2B	45.00	2	3	14.00	5.80	1.88	1.20
3L-45-3B	45.00	3	3	14.00	3.90	1.88	1.20
3L-45-4B	45.00	4	3	14.00	2.90	1.88	1.20
3L-60-2B	61.00	2	3	14.00	5.80	2.50	1.20
3L-60-3B	61.00	3	3	14.00	3.90	2.50	1.20
3L-60-4B	61.00	4	3	14.00	2.90	2.50	1.20
3L-75-2B	75.00	2	3	14.00	5.80	3.10	1.20
3L-75-3B	75.00	3	3	14.00	3.90	3.10	1.20
3L-75-4B	75.00	4	3	14.00	2.90	3.10	1.20
3L-90-2B	90.00	2	3	14.00	5.80	3.75	1.20
3L-90-3B	90.00	3	3	14.00	3.90	3.75	1.20
3L-90-4B	90.00	4	3	14.00	2.90	3.75	1.20
4L-30-4B	30.00	4	4	17.10	3.50	1.25	1.45
4L-30-5B	30.00	5	4	17.10	2.80	1.25	1.45
4L-30-6B	30.00	6	4	17.10	2.40	1.25	1.45
4L-45-3B	45.00	3	4	17.10	4.70	1.88	1.45
4L-45-4B	45.00	4	4	17.10	3.50	1.88	1.45
4L-45-5B	45.00	5	4	17.10	2.80	1.88	1.45
4L-45-6B	45.00	6	4	17.10	2.40	1.88	1.45
4L-60-3B	61.00	3	4	17.10	4.70	2.50	1.45
4L-60-4B	61.00	4	4	17.10	3.50	2.50	1.45
4L-60-5B	61.00	5	4	17.10	2.80	2.50	1.45
4L-60-6B	61.00	6	4	17.10	2.40	2.50	1.45
4L-75-3B	75.00	3	4	17.10	4.70	3.10	1.45
4L-75-4B	75.00	4	4	17.10	3.50	3.10	1.45
4L-75-5B	75.00	5	4	17.10	2.80	3.10	1.45
4L-75-6B	75.00	6	4	17.10	2.40	3.10	1.45
4L-90-3B	90.00	3	4	17.10	4.70	3.75	1.45

Contd...

Bridge ID	Span Length (m)	No. of Box	No of Lane	Bridge Width (m)	Web Spacing (m)	Bridge Depth (m)	Overhang Width (m)
4L-90-4B	90.00	4	4	17.10	3.50	3.75	1.45
4L-90-5B	90.00	5	4	17.10	2.80	3.75	1.45
4L-90-6B	90.00	6	4	17.10	2.40	3.75	1.45

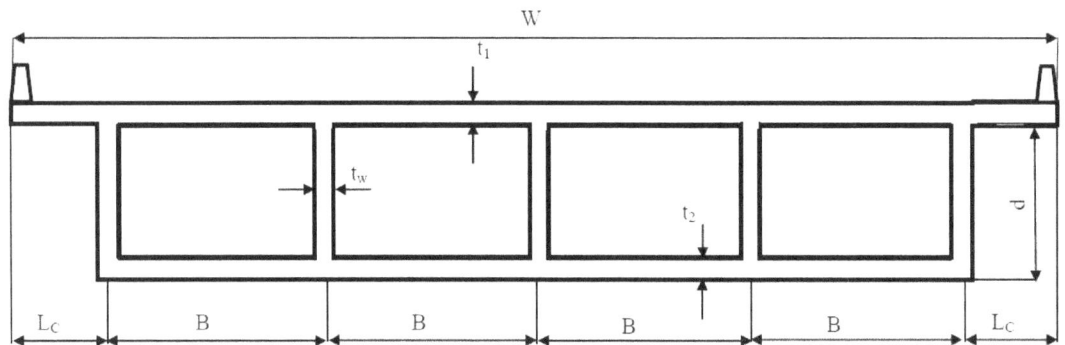

Figure 3.12 Cross Section Symbols for Four-Box Bridge

The width of each design lane loaded should be 3.67 m (10ft.) based on AASHTO LRFD (2008) specifications. In this study, two, three and four lane loads were investigated. The total width of bridge including the traffic road plus two sidewalks would be 9.15 m for two-lane bridges, 14.00 m for three-lane bridges, and 17.10 m for four-lane multicell box-girder bridges.

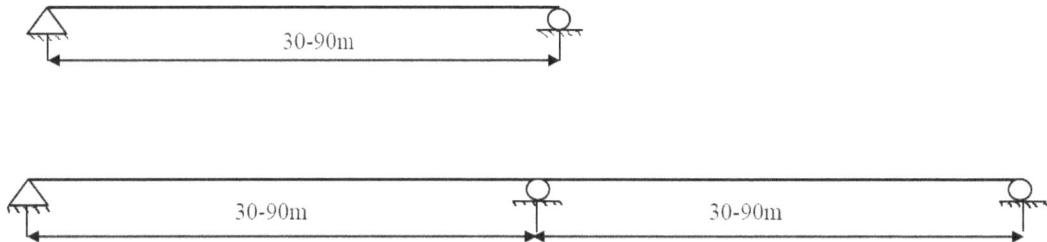

Figure 3.13 Variation of Span Length for Simply Support and Continuous Bridge

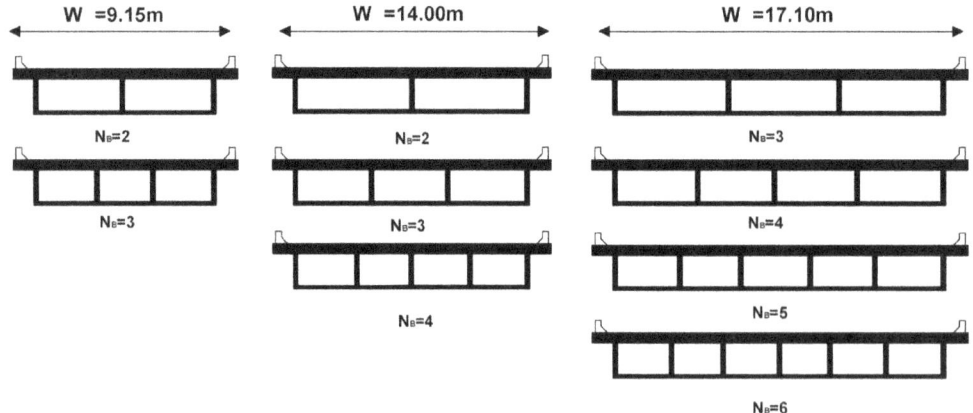

Figure 3.14 Cross Section Configurations with different number of boxes

The number of boxes depends on number of lanes and ranged from two to three for two-lane bridges; two to four for three-lane bridges; and three to six for four-lane bridges. The number of spans, the number of boxes, and variation of skew angle evaluated in this study are indicated in Figure 3.13 through Figure 3.15. To consider the effect of skewness on live load distribution factor and dynamic impact factor of multicell box-girder bridges, the skew angle ranged from 0 to 60° in 15° increment (Khaloo & Mirzabozorg 2003; Ashebo et al. 2007a; Zhang 2008; Theoret et al. 2011).

Hall & Yoo (1999) concluded that span-to-depth ratio between 20 to 30 is the most economical and practical range for box-girder bridges. For concrete multicell box-girder in this study the ratio of 24.0 was chosen based on an empirical examination of a large number of box-girder bridges. The solid end diaphragms were set up inside the box-girder at each support and pier, at the same depth of boxes.

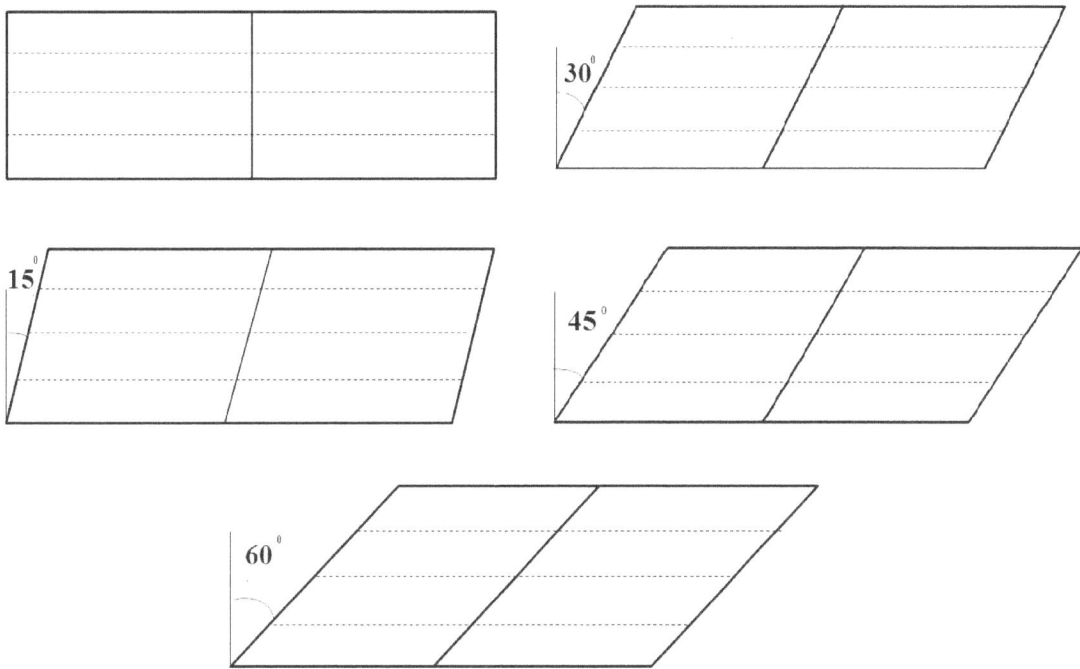

Figure 3.15 Variation of Skew Angle

Samaan (2004) concluded that the effect of upper and bottom slab thickness and web width on live load distribution and dynamic impact factor of box-girder bridges is negligible. Thus, the thickness of web, upper and bottom slab were assumed constant in parametric study. For all prototype bridges employed in parametric study, the modulus of elasticity and passion's ratio were applied as 22.80 Gpa and 0.20, respectively.

In order to develop empirical skew correction factor expressions for bending moment and shear force for simplified Henry's method, in term of LRFD formulas, the bridge database developed by Zhang (2008) were used for this study. The bridge database was constructed using Ontario method (Bakht & Jaeger 1985), named α-θ method. The geometry characteristics of designed bridges were tabulated in Table 3.5. Each of the prototype models presented in Table 3.5 are modeled with skew angle of 0 to 60° in 15° increments. For each databases presented in Table 3.5, five bridge with different skew angle are modeled. For example, for bridge D-1, bridges with skew angle of 0, 15°, 30°, 45° and 60° are modeled.

The main difference between the two sets of bridge database is in the span-to depth ratio (L/d), which is constant (L/d=24) for all bridges in Table 3.4 (Heins 1978), and varied from bridge to bridge in Table 3.5.

Table 3.5 Database for Multicell Box-Girder Bridges

Bridge ID	Span Length (m)	Skew Angle (deg.)*	No. of Boxes (N_B)	Deck Width (m)	Overhang Width (m)	Deck Thick (mm)	Bottom Flange Thick (mm)	Web Thick (mm)	Total Depth (m)
D-1	30		4	11.60	1.13	178	127	203	1.25
D-2	30	0	4	14.60	1.22	203	152	225	1.50
D-3	30	15°	4	17.70	1.52	229	178	250	1.57
D-4	30	30°	4	20.13	1.52	248	208	305	1.65
D-5	45	45°	4	11.60	1.13	178	152	229	1.90
D-6	45	60°	4	14.60	1.22	203	178	250	2.00
D-7	45		4	17.70	1.22	229	208	275	2.12
D-8	45		4	20.13	1.52	248	229	305	2.25
D-9	60		4	11.60	1.13	178	178	254	2.47
D-10	60		4	14.60	1.22	203	208	305	2.60
D-11	60		4	17.70	1.22	229	229	350	2.72
D-12	60		4	20.13	1.52	248	279	375	2.85
D-13	75		4	11.60	1.13	178	203	279	3.00
D-14	75		4	14.60	1.22	203	229	325	3.12
D-15	75		4	17.70	1.22	229	254	375	3.25
D-16	75		4	20.13	1.52	248	279	425	3.40
D-17	90		4	11.60	1.13	178	229	305	3.45
D-18	90		4	14.60	1.22	203	254	350	3.60
D-19	90		4	17.70	1.22	229	279	406	3.75
D-20	90		4	20.13	1.52	248	305	450	3.90

*Each bridge are modeled with skew angle of 0° to 60° in 15° increments

3.11.2 Intermediate Diaphragms Configurations

Based on different recommendations of various bridge design codes on how to use the intermediate diaphragms on bridges, several arrangements diaphragms were considered in this study.

In the first arrangement, the bridges are without any intermediate diaphragm (S-1). In the second arrangement, intermediate diaphragms are parallel to the supporting line (S-2). For this system, according to Louisiana bridge design manual (LADOTD 2002), intermediate diaphragms are applied at one-third and two-third of the span lengths. In the third arrangement, intermediate diaphragms are perpendicular to the longitudinal girders. For this arrangement, three various cases are considered.

In the first case, S-3-1, According to AASHTO (2002), intermediate diaphragm is located at the mid-span of bridges. In the second case, S-3-2, the location of intermediate diaphragms are according to Louisiana bridge design manual, (LADOTD 2002) with two intermediate diaphragms located at the one-third and two-third of span lengths. In the third case, S-3-3, intermediate diaphragms are located at spacing 7.60 m. These various systems of intermediate diaphragms arrangements are listed in Table 3.6 and Figure 3.16.

Table 3.6 Arrangements of Intermediate Diaphragms (IDs)

No. of set	S-1	S-2	S-3-X
Arrangement	No ID	Parallel	perpendicular
Location of ID	-	(1/3 and 2/3) L	X=1 1/2 L X=2 1/3 and 2/3 L X=3 at spacing 7.60 m

Note: The Different Arrangements of IDs Are Shown in Figure 5.5

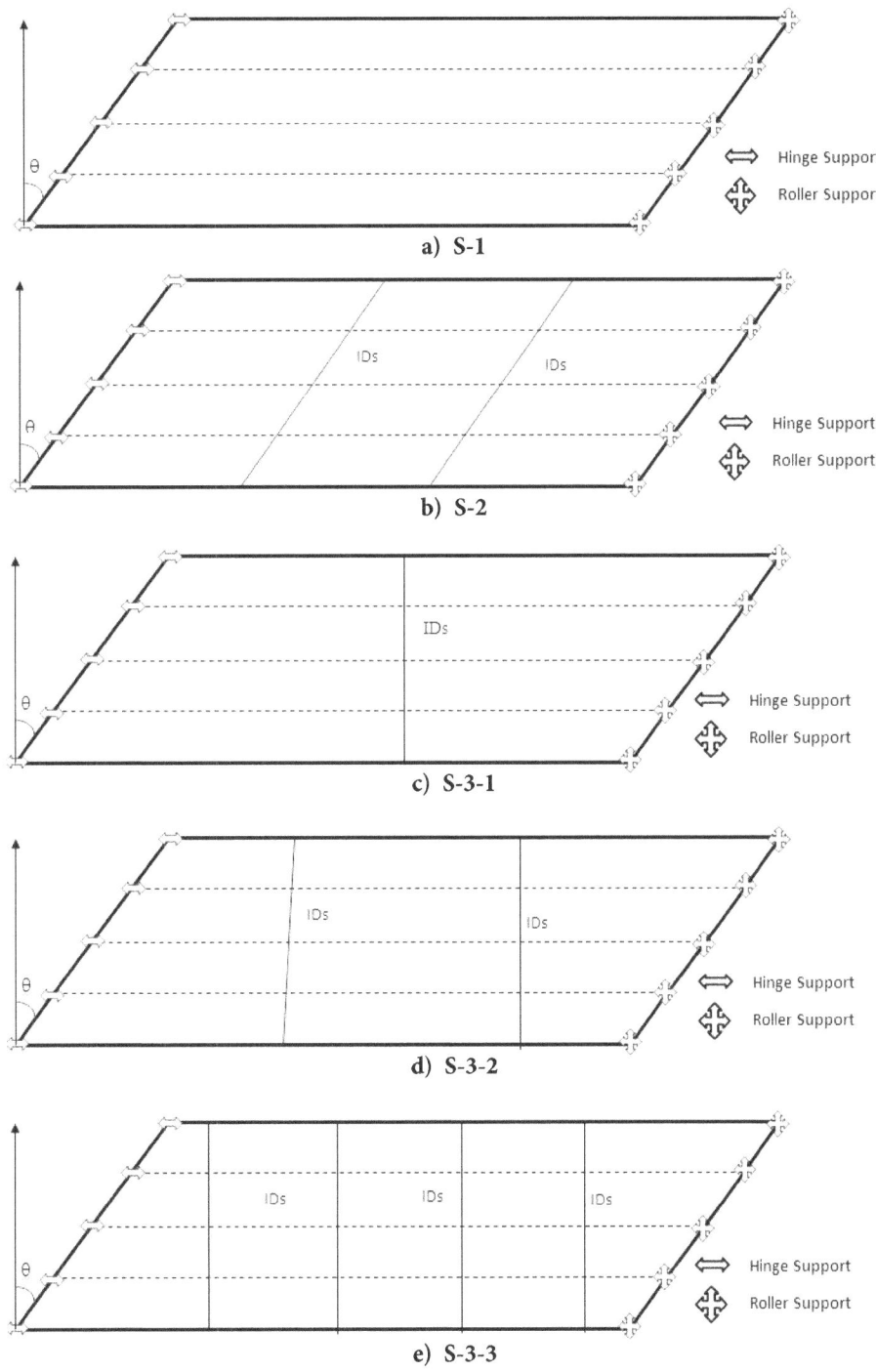

a) S-1

b) S-2

c) S-3-1

d) S-3-2

e) S-3-3

Figure 3.16 Various Arrangements for Intermediate Diaphragms (IDs)

3.11.3 Loading Conditions for Live Load Distribution Factor

The AASHTO LRFD designated vehicle, HL-93, were selected in this study. The designated HL-93 is including design HL-93 truck load (or HS 20-44 from AASHTO standard) plus design lane load or the design tandem plus lane load, whichever governs, was used to obtain maximum bridge actions. The characteristics of truck, tandem and lane load was shown in Figure 3.17.

A case with 90% of two truck spaced a minimum distance of 15.20 m apart in the longitudinal direction plus 90% lane load was used for computing compressive stress, reaction and shear force at the piers. The configuration of live load on longitudinal direction was indicated in Figure 3.18.

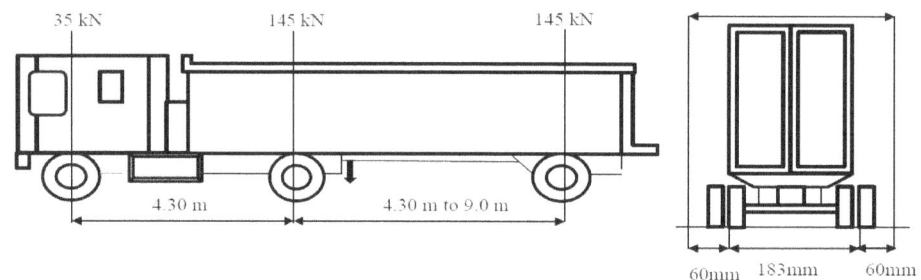

a) HL-93 truck

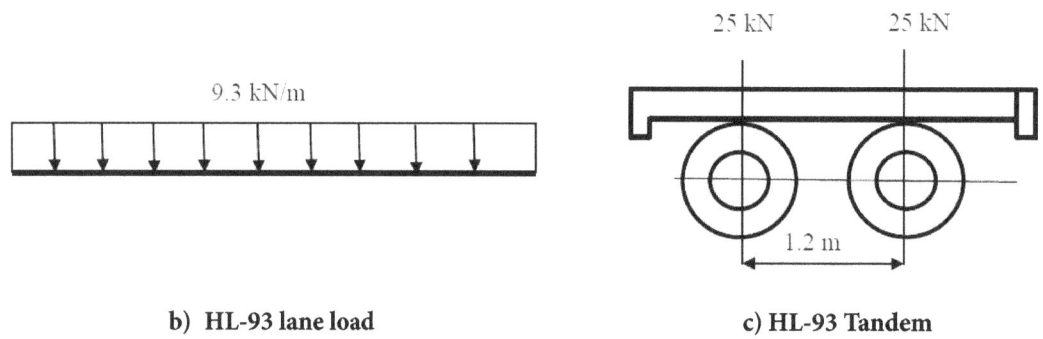

b) HL-93 lane load   c) HL-93 Tandem

Figure 3.17 AASHTO LRFD Live Loading

AASHTO LRFD vehicle were applied according to number of lanes so that for two lane load bridges, one and two lane loaded case were investigated. For three lane loaded bridges, one, two, and three lane loaded was examined. For four lane bridges, two, three and four lane loaded were considered.

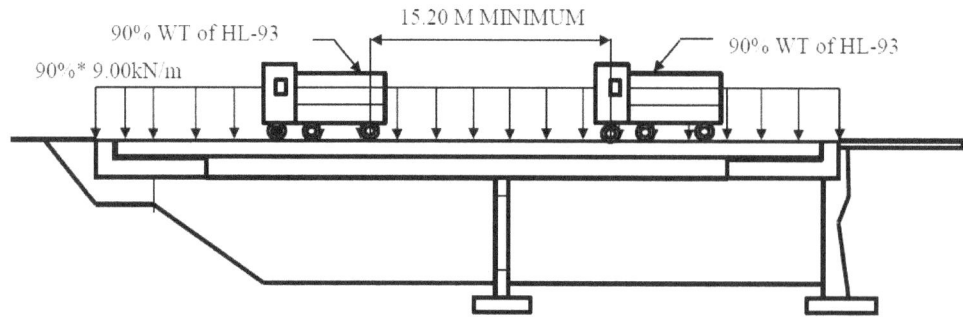

Figure 3.18 Live Loading Position for Maximum Reaction and Compressive Stress

By employing the finite element technique for three-dimensional model of bridges, the maximum responses were obtained by placing the vehicle loads at a distance of 60 mm from the corner of the curbs, and then the truck moved all together in the longitudinal direction. Then, the multiple presence factors of 1.00, 0.85 and 0.65 for two, three and four lane loading, respectively, were applied based on AASHTO LRFD (2008) specifications. The adjacent wheel line of two trucks was placed 1.22 m apart from each other. The location of truck in the transverse direction of bridge is shown in Figure 3.19.

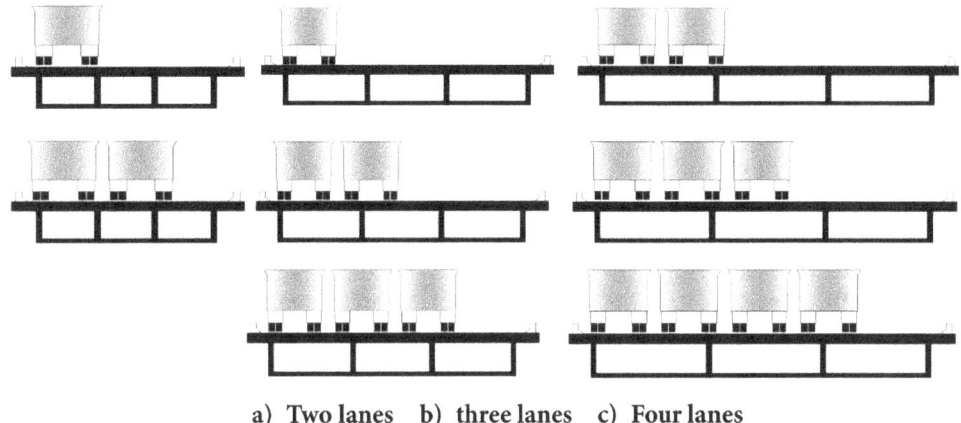

a) Two lanes   b) three lanes   c) Four lanes

Figure 3.19 HL-93 Loading Cases in the Transverse Direction of the Bridges

3.11.4 Parametric Study for Dynamic Impact Factor

An extensive study first performed to evaluate the effect of various parameters such as the truck velocity on dynamic bridge-vehicle interactions. This will be discussed later in Chapter VII.

3.11.4.1 Loading condition for dynamic impact factors

Only three-axle HL-93 truck (see Figure 3.20) was used to evaluate the dynamic bridge-vehicle interaction. Since the mass of the prototype bridges in this study were more than 10 times the mass of vehicle, the effect of vehicle mass were therefore neglected in the first element analysis (Senthilvasan et al. 1997). Multiple presence factor were not apply for parametric study. It was assumed that truck moved smoothly with no slippage and at a constant speed on the bridge.

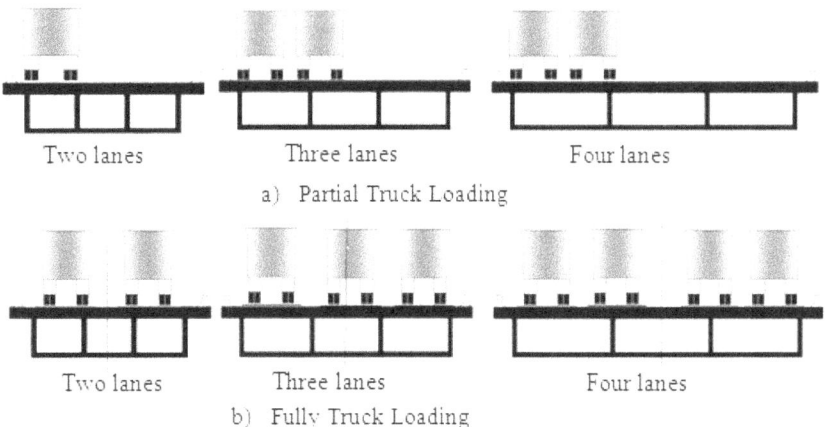

Figure 3.20 Truck Loading Case in the Transverse Direction for Dynamic Analysis of Bridges

Two set of loading condition named fully loaded and partially loaded were investigated in the transvers direction for each bridge to evaluate the dynamic bridge responses. In the partially loading condition, the truck positioned at a distance of 600 mm from the corner of curb of bridge, however, in the fully loading condition, all trucks positioned in the center of a lane. The truck position on transvers direction of a three box-girder bridge in both fully and partial loading conditions are shown in Figure 3.20.

3.12 Summary

In this Chapter, two set of database were developed to evaluate the dynamic and static behaviour of multicell box-girder bridges. The first set was constructed based on Heins's method, with span-to-depth ratio of 24 and constant values for deck thickness and bottom flange thickness of prototype bridges. Second set of database was established upon Ontario method using α-θ technique. This database will be used to develop proposed equations for skew correction factor of live load distribution factor from simplified Henry's method, in the same term as LRFD formulas for skew correction factors. To evaluate the effect of concrete solid intermediate diaphragms on distribution of live load, three various configurations were depicted, named without, parallel to abutment, and perpendicular to girder intermediate diaphragms. In order to consider the effectiveness of number of intermediate diaphragms on bridge responses, the recommended of various codes on distance between IDs were described. The LRFD live loading vehicle, HL-93 including designated HL-93 truck, tandem, and uniform lane load were selected for parametric studies. The truck position on longitudinal and transvers directions, and number of truck loads for static analysis of bridges were depicted in this chapter. Only truck loads were used for dynamic analysis of bridge. Fully and partially truck loading condition are considered to obtain the maximum bridge responses for dynamic impact factor.

DYNAMIC IMPACT FACTOR

In this chapter the dynamic impact factors for continuous straight and skewed multicell box-girder bridges is discussed. The key parameters that influence the dynamic bridge-vehicle response of skew continuous concrete box girder bridges were studied. Methodology of the vehicle idealization, loading positions and the vehicle speed were presented in Chapter IV. Forced vibration analysis methods are described and compared. Stability and preciseness of the analytical technique were discussed.

A comparative study between the mode superposition and direct integration methods was conducted to select the most effective dynamic analysis method in investigating the impact factors. A sensitive parametric study was then performed to evaluate the effects of the various parameters on the dynamic impact factors of multicell box-girder bridges. Results obtained from the parametric study of 180 continuous skew and straight multicell box-girder bridges (Table 5.1) were analyzed. The effects of number of lanes, number of boxes, bridge span length and skew angle, vehicle speed were discussed.

Then empirical expressions for the dynamic impact factors or dynamic amplification factors were developed for design of multicell box-girder bridges. The proposed expressions were compared with the current equations recommended in the AASHTO (2002) standard specification, AASHTO LRFD (2008) specifications (2008), CHBDC (2000).

4.1   Comparison of Mode Superposition Method and Direct Integration Method

Two analysis techniques for evaluation of dynamic bridge-vehicle interaction were introduced in Chapter IV, called superposition method and direct integration method. A three-box continuous bridge with span length of 60 m was analyzed using two described analyzing techniques.

The value of dynamic bridge actions was determined using the mode superposition method by taking into account first 300 mode shapes (Samaan 2004). The results shown at influence line for bending moment and shear force then were compared to values obtained from the direct integration method as shown in Figure 4.1 and Figure 4.2, respectively. Obviously, the maximum bending moment values from direct integration method was slightly higher than those from mode superposition technique. In addition, due to necessity of performing a modal analysis to derive the natural frequencies of bridge and then clustering the obtained frequencies, the time required for solution using mode superposition technique is not lower than the direct integration method. Thus, it was concluded that the direct integration method was the most suitable method to obtain accurate results and reasonable computation time in dynamic analysis of structures.

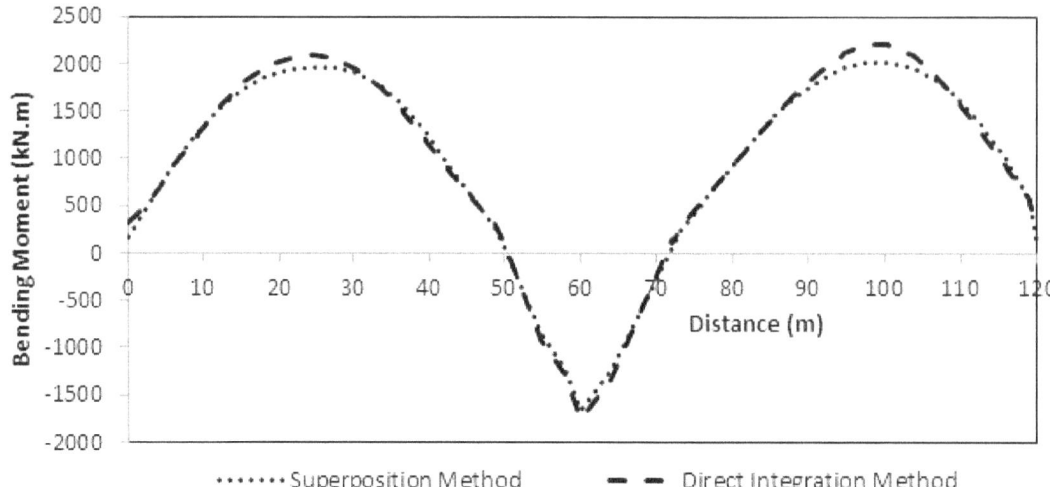

Figure 4.1 Comparisons between Direct Integration and Mode Superposition Methods for Positive Bending Moment

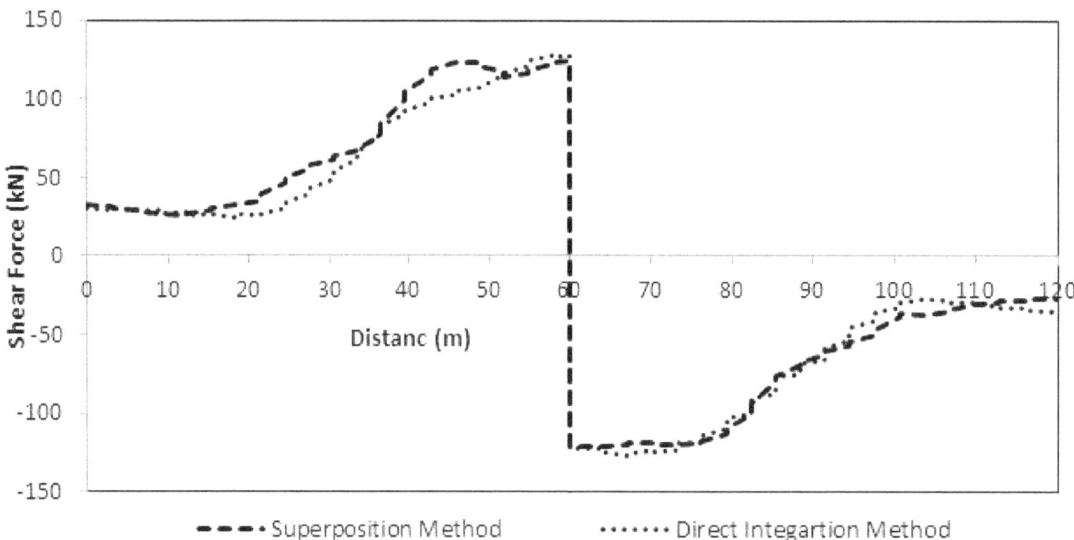

Figure 4.2 Comparisons between Direct Integration and Mode Superposition Methods for Shear Force

4.2   Damping Effect on Dynamic Response of Bridge

It is clear that the magnitude of oscillation of bridge reduces until the fluctuation stop because of waste of energy. The absorbed energy in structure is called damping. By employing the direct integration method available in SAP2000 software, called Hilber-Hughes-Taylor (Hilber & Hughes 1978), a hypothetical damping is provided which is fully analytical. The α-operator presented in this method introduced damping of structure. This α-operator at any values between 0 to –0.33, for minimum and maximum damping, respectively.

The three-box bridges with span length of 60 were analyzed to consider the effect of damping on dynamic bridge-vehicle interaction responses. The α-operator represented of damping, of 0, –0.05, –0.10 and –0.33 were selected for this study. The results of damping effect on dynamic responses of bending moment and shear force are shown in Figure 4.3 and Figure 4.4, respectively. It can be observed that the maximum results of all bridge actions were only slightly different, when damping increased. Thus, the damping of zero was applied for all prototype bridge in this study.

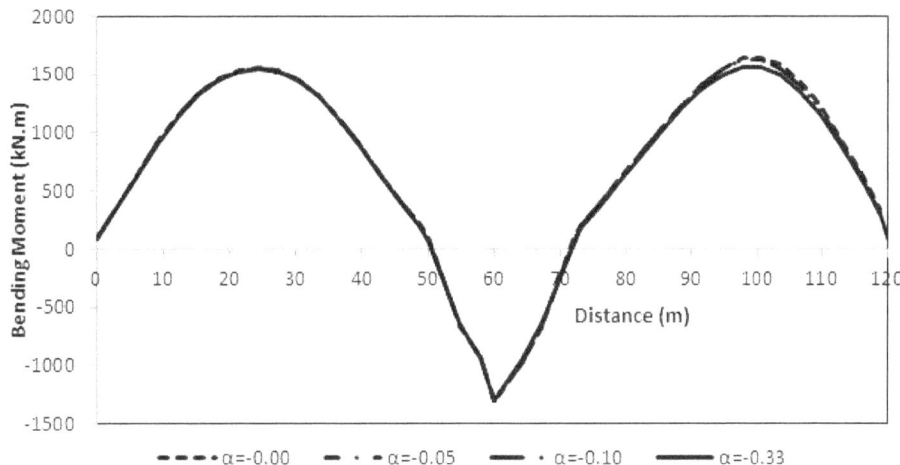

Figure 4.3 Damping Effect on Dynamic Bending Moment of a 60 m, Three-Box Bridge

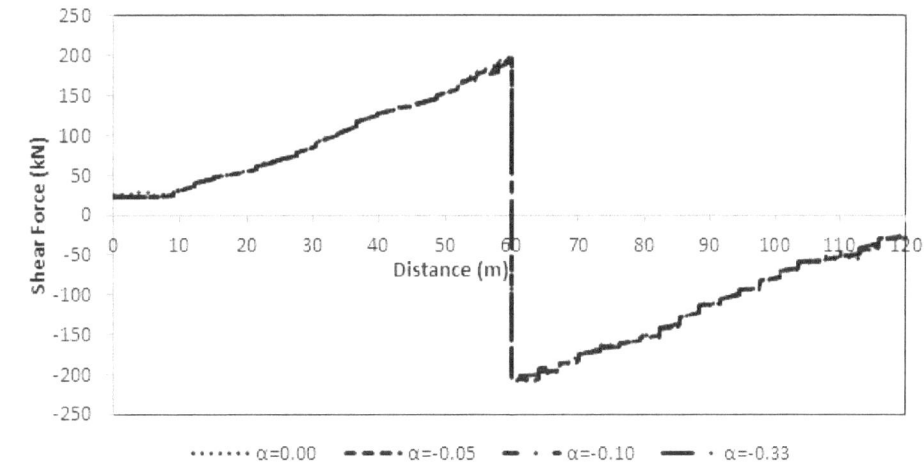

Figure 4.4 Damping Effect on Dynamic Shear Force of a 60 m, Three Box Bridge

4.3   Parametric Study on Dynamic Impact Factor

The literature recommended that the dynamic impact factor of live loads moving on a bridge superstructure depends upon a large number of variables. In this research the following key parameters were chosen to have the highest effect on the dynamic bridge-vehicle interaction; namely: 1) Number of lane loaded; 2) Number of boxes; 3) span length; 4) skew angle; and 5) vehicle speed. In the parametric study of live load distribution factor, most of these factors were thoroughly studied. In the following section, results were indicated in graphs and finally two set of expression for dynamic impact factors were proposed.

4.3.1 Effect of Number of Lanes

Number of loaded lanes investigated herein was two, three and four. Since the number of lanes was a representation of the variation of the bridge width, this parameter was evaluated as a main parameter in the bridge geometry and consequently, one that importantly affects the bridge responses subjected to dynamic loads. Dynamic impact factor for bending moment, vertical shear and reaction of three-boxes bridge with two, three and four lane loads is plotted in Figure 4.5 through Figure 4.7, respectively.

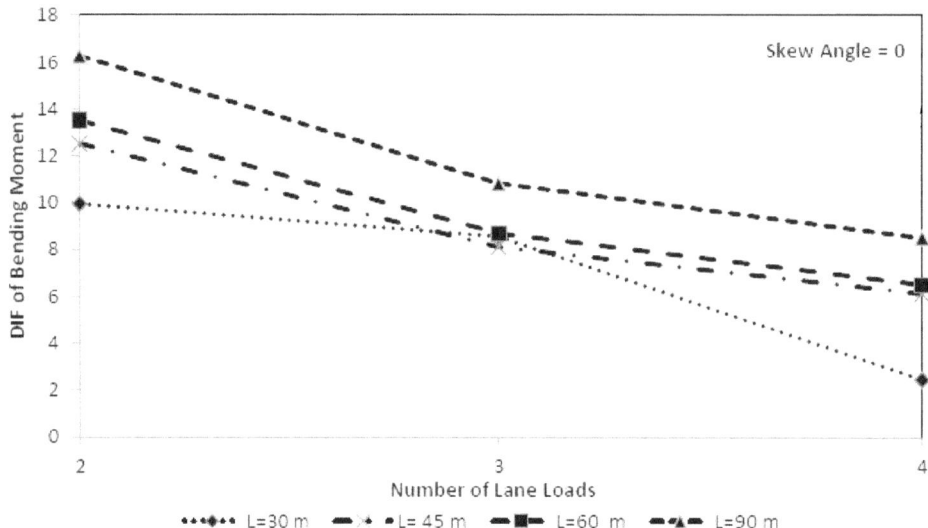

Figure 4.5 Dynamic Impact Factor of Bending Moment versus Number of Lane Loads for Three-Box Bridges

The graphs indicate that the dynamic impact factor of bending moment, shear force and reaction do not follow a certain trend for bridges with increase in the number of lane loaded. Similar observations were revealed for dynamic impact factors for deflection, but did not present due to repetition. Based on the above results, it can be concluded that the number of lanes, and hence the bridge width, does not have a patterned influence on the dynamic impact factors for continuous multicell box girder bridges. This finding was fully compatible with North American codes (AASHTO 2004).

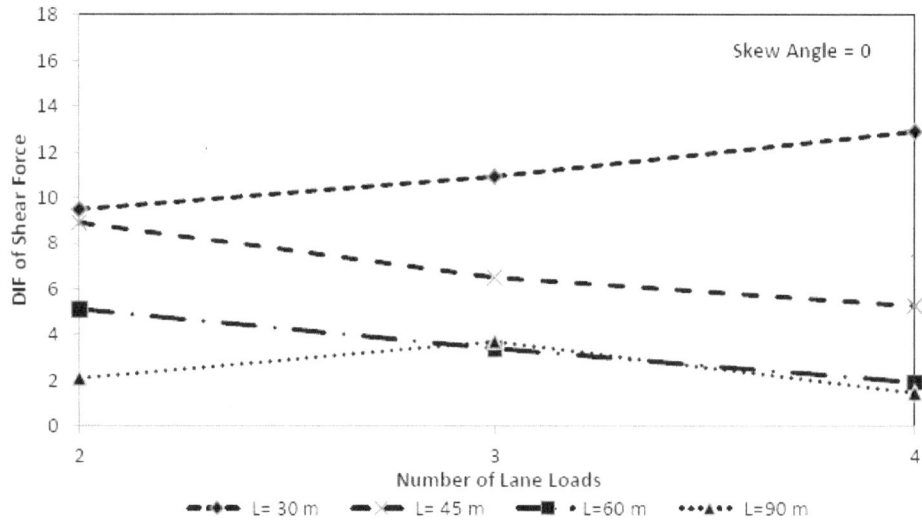

Figure 4.6 Dynamic Impact Factor of Shear Force versus Number of Lane Loads for Three-Box Bridges

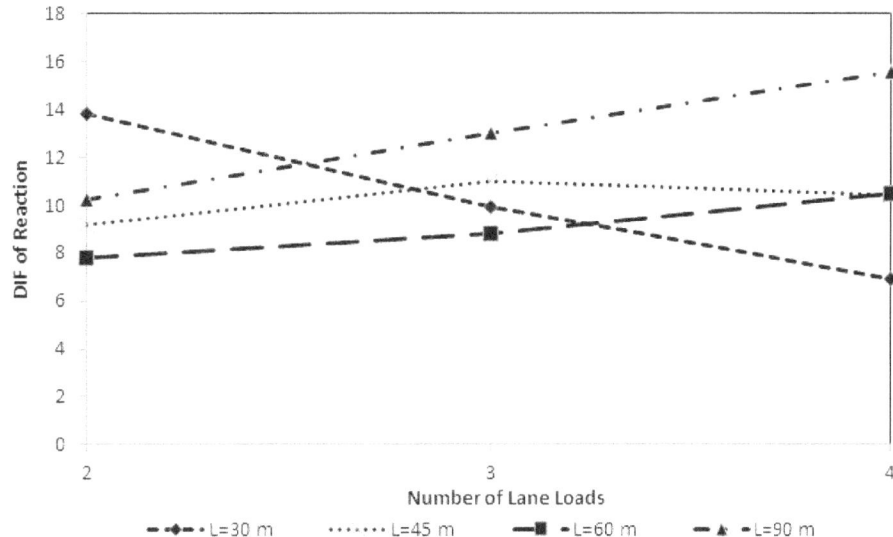

Figure 4.7 Dynamic Impact Factor of Reaction versus Number of Lane Loads for Three-Box Bridges

4.3.2   Effect of Number of Boxes

The effect of the number of boxes on the dynamic impact factors for straight bridges were investigated in this section. The span length of 30 m, 45 m and 60 m bridge were selected with number of boxes ranged from three to six. The relationship between the number of box and dynamic impact factor of bending moment, shear force and reaction for straight bridges is plotted in Figure 4.8 through Figure 4.10.

It can be observed that the dynamic impact factor for bending moment generally increased with increasing number of boxes. The variation of dynamic impact factor for reaction and shear force do not follow a clear systematic trend with changing the number of boxes. The same results were also obtained for dynamic impact factor for maximum deflection. Thus, the variation of the impact factor with number of boxes was not considered in developing the proposed equation. It was also compatible with the observation of AASHTO (2002) standard specification and Zhang et al. (2003).

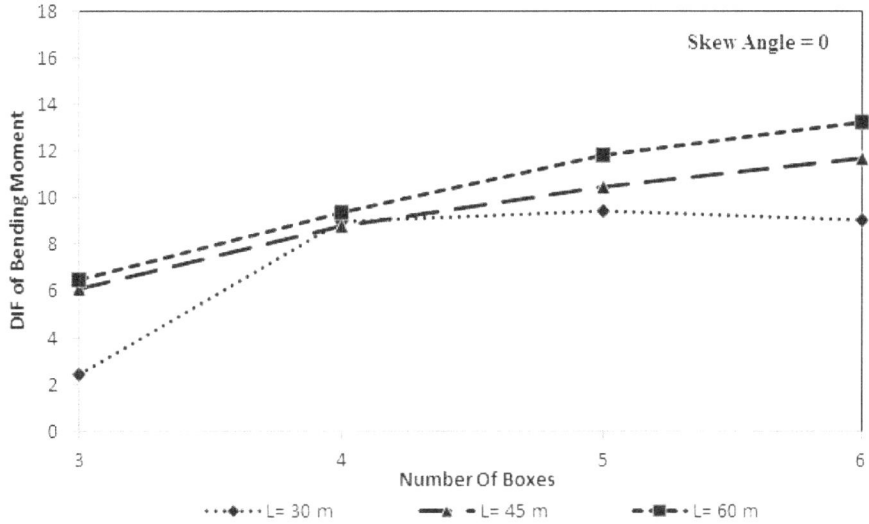

Figure 4.8 Bending Moment Dynamic Impact Factor vs. Number of Boxes

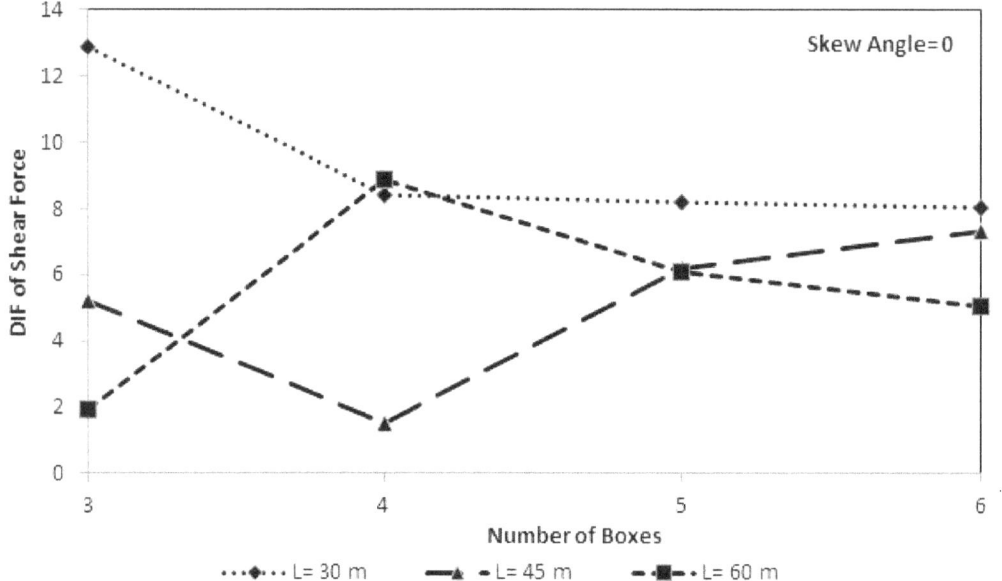

Figure 4.9 Shear Force Dynamic Impact Factor vs. Number of Boxes

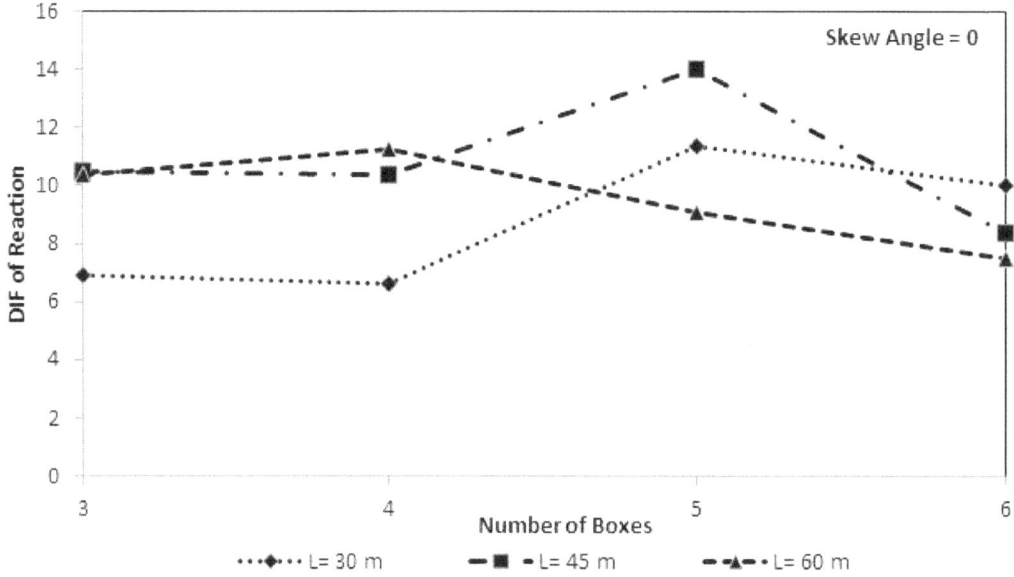

Figure 4.10 Reaction Dynamic Impact Factor vs. Number of Boxes

4.3.3 Effect of Span Length

The relationship between the bridge span length and the dynamic impact factors for bending moment for two, three and four lane loads bridges with three-boxes is shown in Figure 4.11. It can be observed that the dynamic impact factor of bending moment, in general, increased when span length increased from 30 m span to 90 m span.

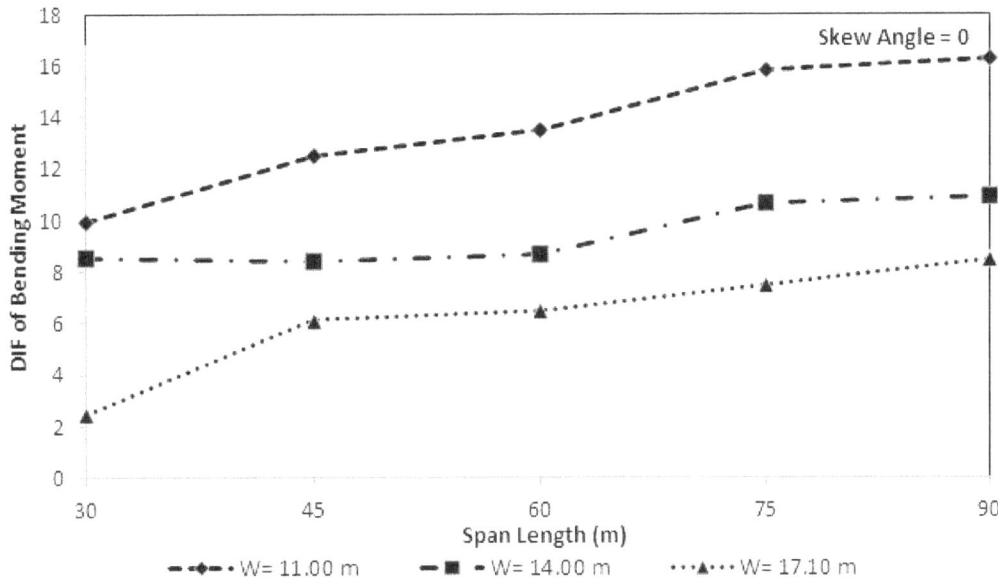

Figure 4.11 Bending Moment Dynamic Impact Factor vs. Span Length

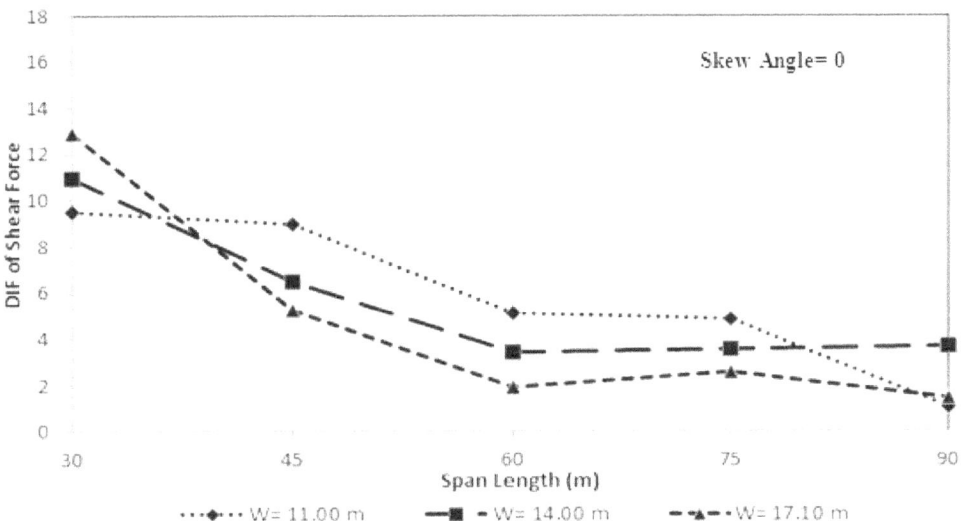

Figure 4.12 Shear Dynamic Impact Factor vs. Span Length

Figure 4.12 shows the effect of span length on dynamic impact factor of shear force of selected bridges. It can be observed that the dynamic impact factor of shear force decreased when span length changed from 30 m to 75m, but the dynamic impact factor slightly increased from 60 m span to 90 m span of bridges with three lane loads. It is difficult to find unique equations to predict this variation correctly, hence the upper bound envelope equations were determined to obtain a conservative value of dynamic impact factor of bridges.

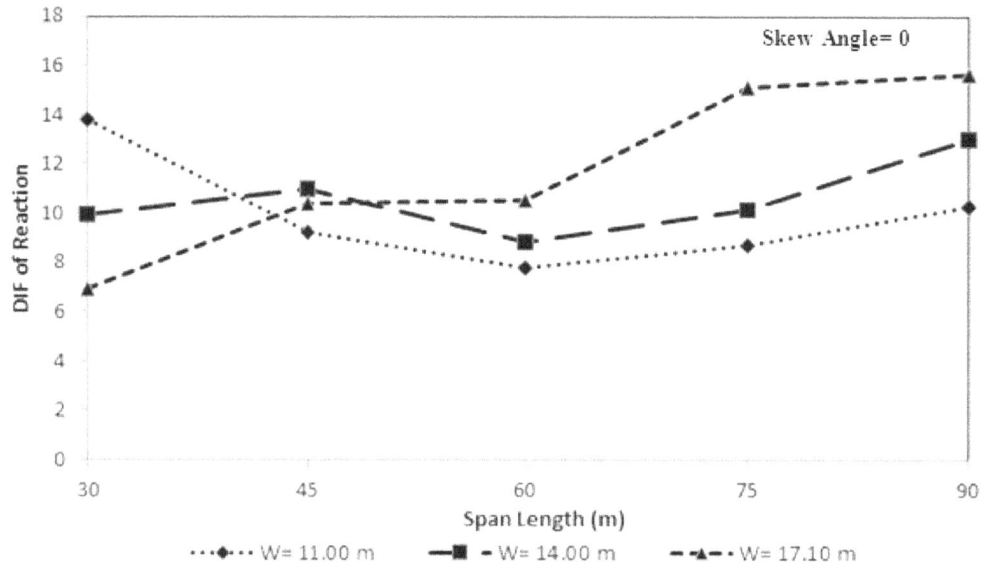

Figure 4.13 Reaction Dynamic Impact Factor vs. Span Length

From Figure 4.13, it was concluded that the dynamic impact factor of reaction did not follow any clear trend for bridges with 30 m span to 60 m span, but the dynamic impact factor for reaction increased when span length increased from 60 m to 90 m. The main objective of this section was to develop proposed equation for dynamic impact factor in term of AASHTO standard specifications, so the effect of span length was considered in developing the dynamic impact factor equation.

4.3.4 Effect of Skew Angle

Although, a large number of studies have been carried out on distribution of vehicle loads on bridges but only a few investigations have been done to understand the effect of skew angle on the dynamic response of bridges. Ashebo et al. (2007 a, b) revealed that the AASHTO LRFD specification predicts conservative value for the dynamic impact factor (DIF) of skewed bridges.

The effect of skew angle on dynamic impact factor of skew multicell box-girder bridge, with span length of 30 m and 60 m are shown in Figure 4.14 and Figure 4.15.

In general, the impact factor of bending moment decreased when skew angle increased. The dynamic impact factor for reaction and shear force did not follow a clear trend when skew angle increased. It can be observed that the dynamic impact factor for shear and reaction highly decreased or increased when skew angle exceed from 45°. It is due to increase the reaction force at acute corner of skewed bridges.

The same variation can be observed for other prototype bridges. Thus, neglecting the effect of skewness on dynamic impact factor may results in highly conservative or unsafe value of dynamic impact factors. For instance, the dynamic impact factor for reaction for bridge with skew angle of 60° was almost 60% higher than corresponding straight bridge.

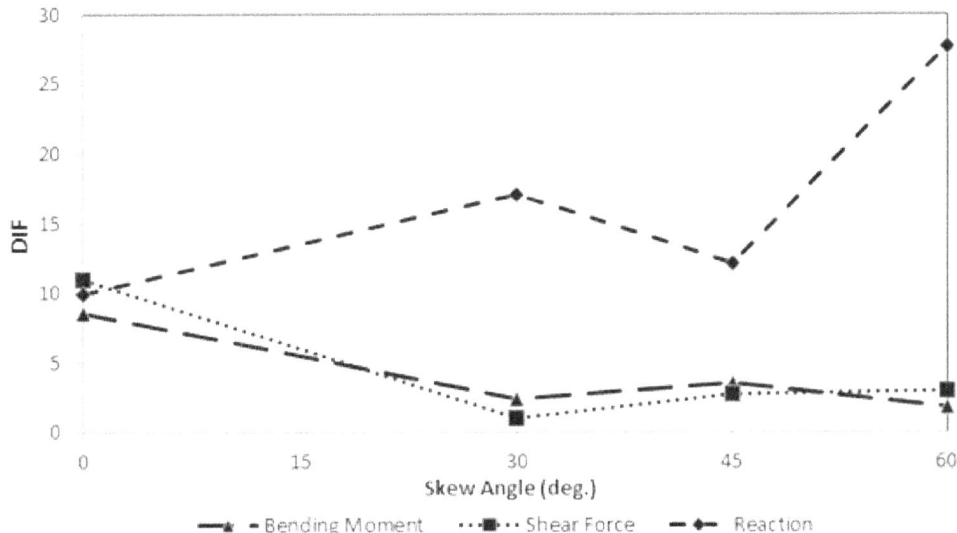

Figure 4.14 Dynamic Impact Factor versus Skew Angle for 30 m span bridge

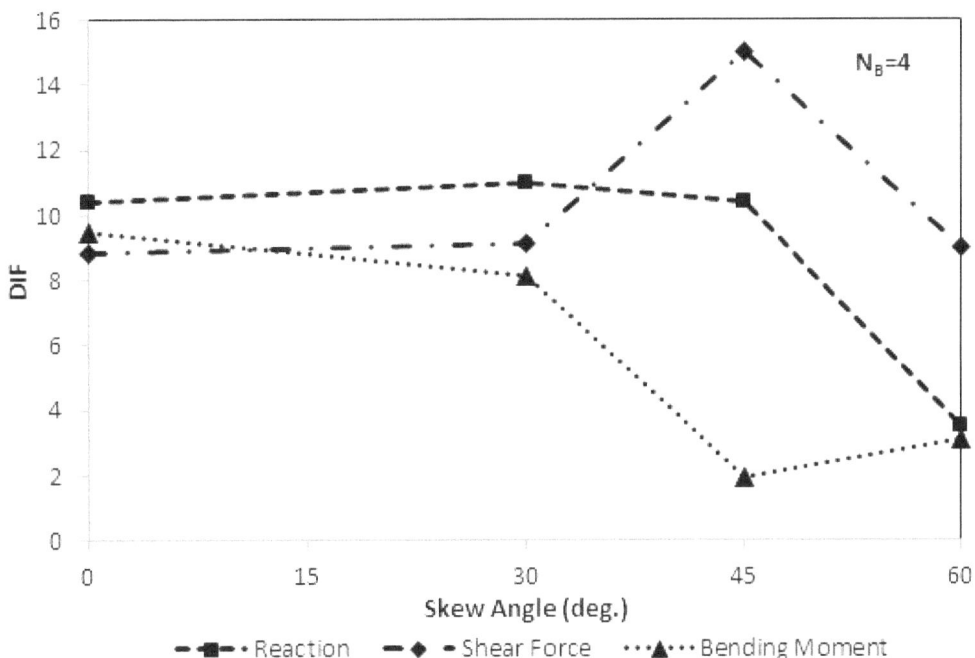

Figure 4.15 Dynamic Impact Factor vs. Skew Angle

4.3.5 Effect of Vehicle Speed

In order to select a suitable vehicle speed for the multicell box-girder bridges, the effect of vehicle speed was investigated for three-box bridges with 60 m span length. Figure 4.16 indicates that when speed increased from 80 km/h (Kilometre per hours) to 140 km/h, the dynamic impact factor for reaction increased by almost 54%. Dynamic impact factor for bending moment increased from 4.3 to 5.4 by up to 20%.

It was observed that the increase in the vehicle speed to 140 km/h produced a slight higher reaction at the internal supports. In addition, the vehicle speed had a slight effect on dynamic impact factor on shear force.

In the light of above discussion, the vehicle speed of 120 km/h which was slightly more than allowance speed accepted for most highways around the world was selected for this research. It was because those drivers often travel with speed higher than the posted road vehicle speed.

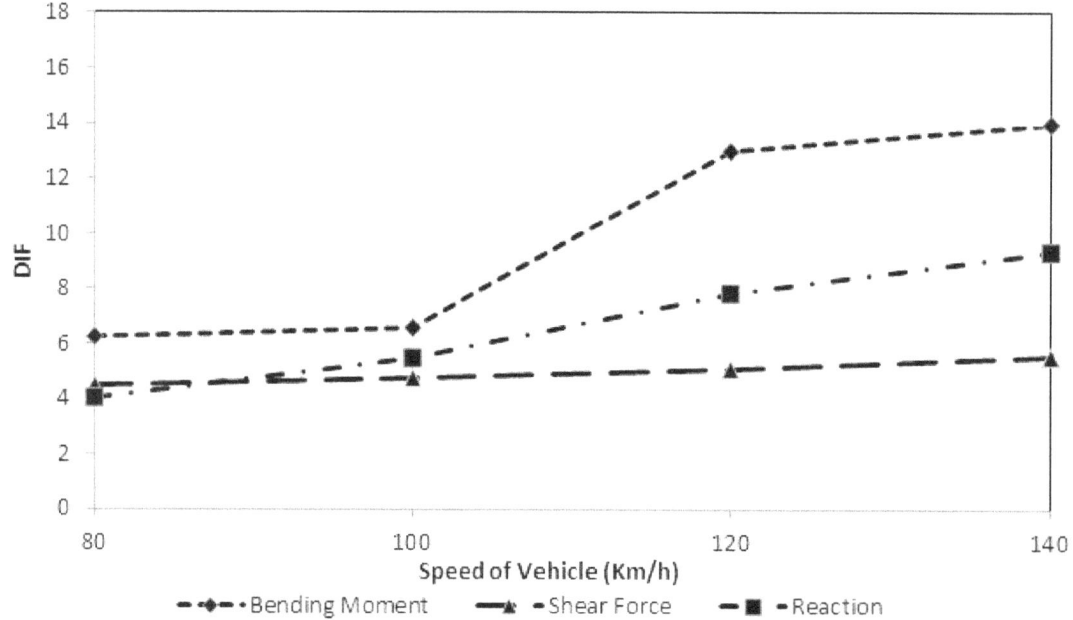

Figure 4.16 Dynamic Impact Factor vs. Vehicle Speed for a Three-Box Bridge with Span length of 60 m

4.4   The First Fundamental Frequency

Previous studies revealed that the first fundamental frequency that is an independent characteristic of bridge is the governing frequency for all straight bridges (Zhang et al. 2003). The first fundamental frequency (the first natural frequency) is only a function of geometric properties of the bridge. The variations of these parameters versus skew angle are plotted in Figure 4.17.

It can be observed that the magnitude of the parameter increased with skewness, however these variations was strongly depended on bridge span lengths. Meanwhile, a decrease in fundamental frequency in longer bridges (Figure 7.17) was probably occurred because the torsional modes were governing in longer bridges with higher skewness (Ashebo et al. 2007a, b); while the bending modes were dominated by the declining bridge skew angle. Since the natural frequency of each bridge is unique, many other researchers and specifications adopted it as an index of the dynamic behaviour of bridges under truck loading (CHBDC 2000; Zhang et al. 2003).

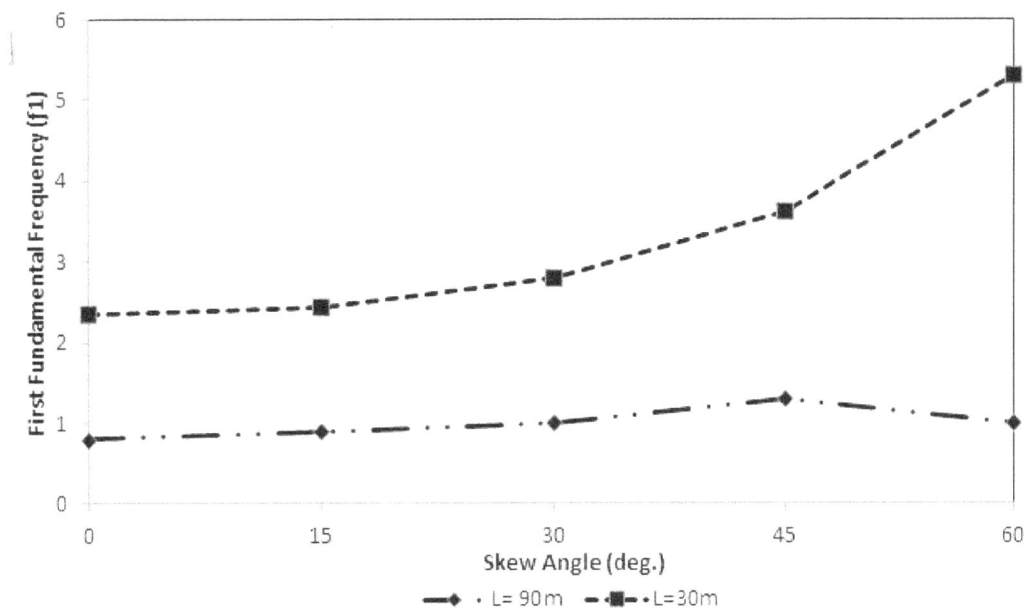

Figure 4.17 First Fundamental Frequencies vs. Skew Angle

4.5 Proposed Equations for Impact Factor

In the previous sections, the effects of main parameters on the dynamic impact factors of skew and straight continuous multicell box-girder bridges were investigated employing the finite element computer program SAP2000. The maximum selected design speed was 120km/h and the fully and partially loading condition was applied for all bridges considered in this parametric study.

Impact factor values were plotted versus the number of lane loaded, number of bridge boxes, span length of bridge, skew angle of bridges and vehicle speed. In some cases, the dynamic impact factors have indicated an obvious pattern with regard to a parameter. However, in many cases no pattern was observed. As expected, dynamic impact factors for bending moment, deflection, reaction forces and shear forces were different by each of the above parameters. This is because of the complex structural dynamic behaviour of continuous multicell box-girder bridges.

Previous study often related the span length of a bridge to its dynamic impact factor (Sennah et al. 2004; Samaan et al. 2007). Also, bridge first fundamental frequency was considered as the most critical factor in predicting dynamic impact factors in most codes (OHBDC 1983; Zhang et al. 2003). Thus, in the following the dynamic impact facts of expressions were developed as a function of span length and first fundamental frequency.

4.5.1 Dynamic Impact Factor as a Function of Fundamental Frequency

The dynamic impact factor responses for all bridge analysis of straight and skew multicell box-girder bridges subjected to moving loads for bending moment, shear force, reaction and also maximum deflection were plotted versus fundamental frequencies of bridges as indicated in Figure 4.18 through Figure 4.20. The obtained results for dynamic impact factor from current bridge design codes including AASHTO LRFD (2008), CHBDC (2000) and OHBDC (1983) are indicated in the graphs for comparing with proposed equations. The expressions for current codes are discussed. Upper bound dynamic impact factor results were used to derive the following equations:

1) Dynamic impact factor for bending moment

$$DIF_m = -2.06 \times f1 + 19.83 \qquad (4.1)$$

2) Dynamic impact factor for shear force

$$DIF_V = -1.1 \times f1 + 26.55 \qquad (4.2)$$

3) Dynamic impact factor for reaction at piers

$$DIF_R = 1.87 \times f1 + 21.74 \qquad (4.3)$$

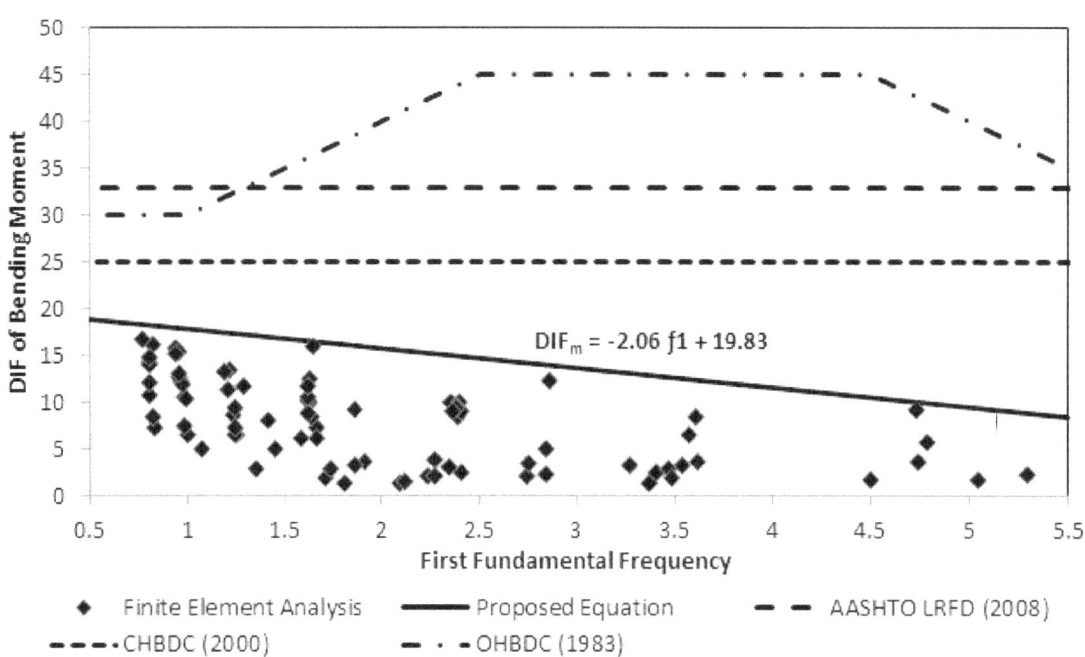

Figure 4.18 Dynamic Impact Factor for Bending Moment as Function of Fundamental Frequency

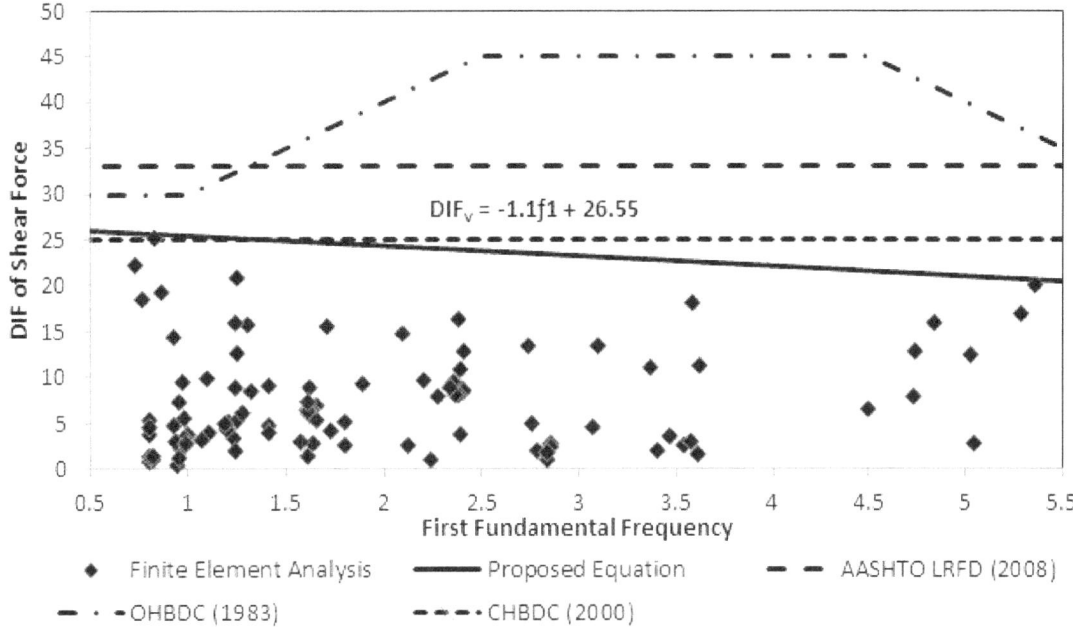

Figure 4.19 Dynamic Impact Factor for Shear Force as Function of Fundamental Frequency

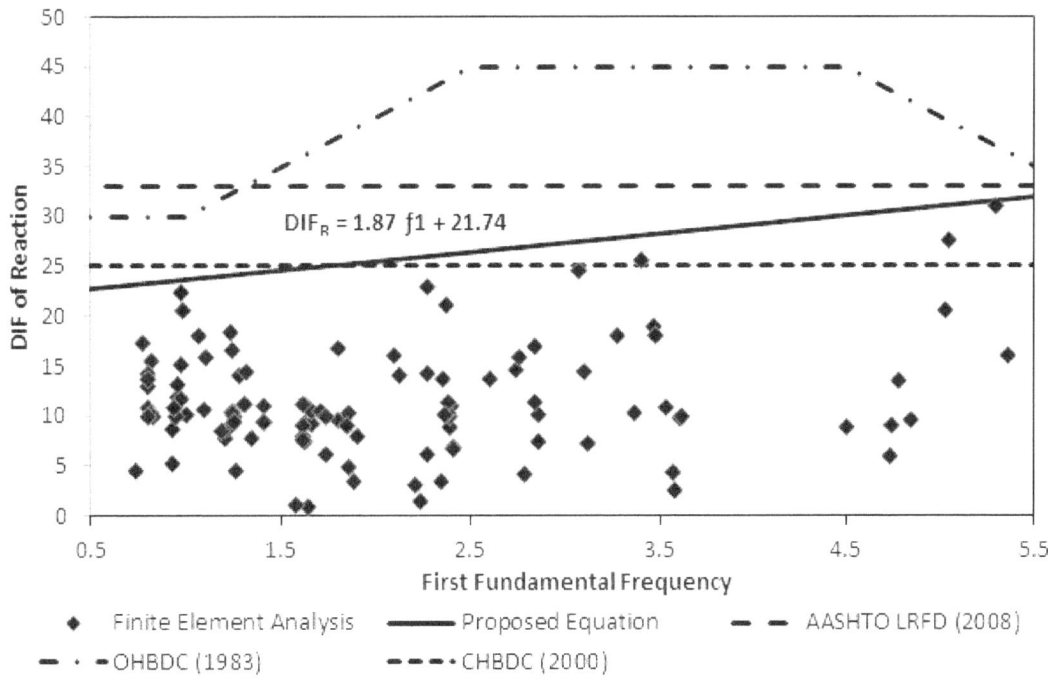

Figure 4.20 Dynamic Impact Factor for Reaction as Function of Fundamental Frequency

It can be observed that current specification estimate highly conservative values for dynamic impact factor of bending moment. AASHTO LRFD (2008) specification and Ontario Highway and Bridge Design Code (OHBDC 1983) estimate conservative values for dynamic impact factor of shear force and reaction. The dynamic impact factor results for shear force obtained from CHBDC (2000) were suitably reasonable, although obtained unsafe values for dynamic impact factor of reaction.

4.5.2 Dynamic Impact Factor as Function of Span Length

The dynamic impact factor values are shown as function of span length of multicell box-girder bridges in Figure 4.21 through Figure 4.23. The dynamic impact factor values from available formulas presented in AASHTO (2002), AASHTO LRFD (2008), CHBDC (2000) and JRA (1996) bridge specifications were also determined for comparison with proposed equations. The expressions for current codes are discussed. Proposed equations for dynamic impact factor as function of span length were derived from upper bound results of finite element results as following:

1) Dynamic impact factor for bending moment

$$\mathrm{DIF}_m = 0.127 \times \mathrm{L} + 9.40 \quad \mathrm{L} \le 60 \text{ m} \qquad (4.4A)$$

$$\mathrm{DIF}_m = 17.0 \quad 60 \text{ m} \le \mathrm{L} \le 90 \text{ m} \qquad (4.4B)$$

2) Dynamic impact factor for shear force

$$\mathrm{DIF}_V = 0.09 \times \mathrm{L} + 17.90 \qquad (7.5)$$

3) Dynamic impact factor for reaction at piers

$$\mathrm{DIF}_R = -0.142 \times \mathrm{L} + 35.75 \qquad (7.6)$$

As mentioned earlier, the AASHTO LRFD specification predicts highly conservative values for dynamic impact factor of bridges. The AASHTO standard specification (AASHTO, 2002) and JRA (1996) estimate conservative results for dynamic impact factor for bending moment, when span length is less than 60 m, and obtain unsafe values for longer bridges. The average discrepancy between CHBDC (2000) code and proposed equations was 15% which indicated a relatively reasonable method for determine the dynamic impact factor of shear force. in addition, AASHTO (2002) standard specifications and JRA (1996) obtain highly unsafe values for both shear force and reaction at piers.

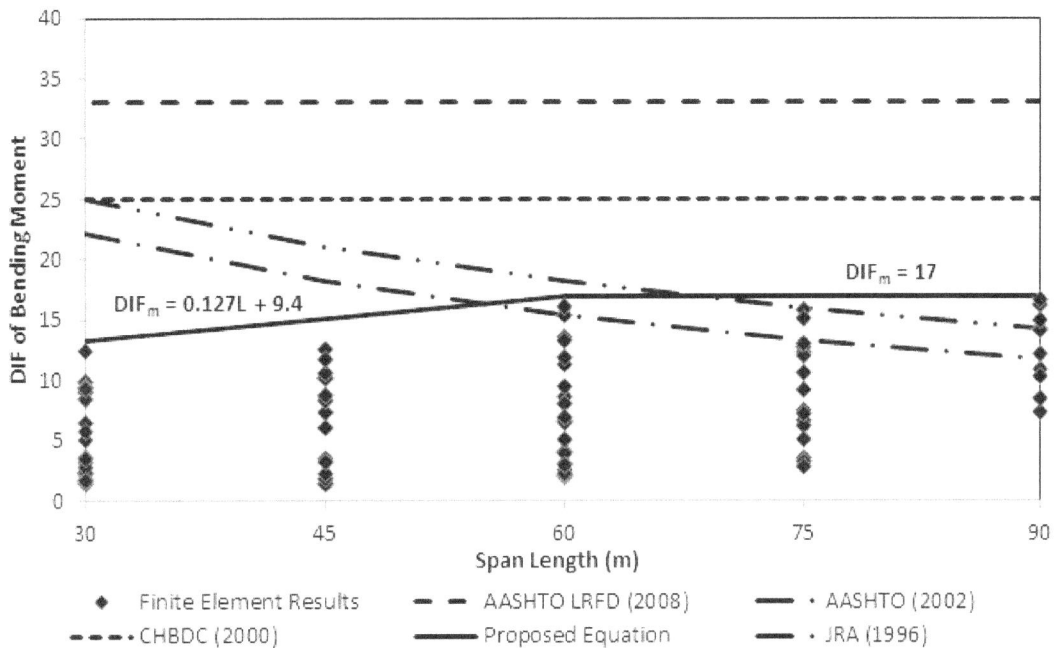

Figure 4.21 Dynamic Impact Factor for Bending Moment as Function of Span Length

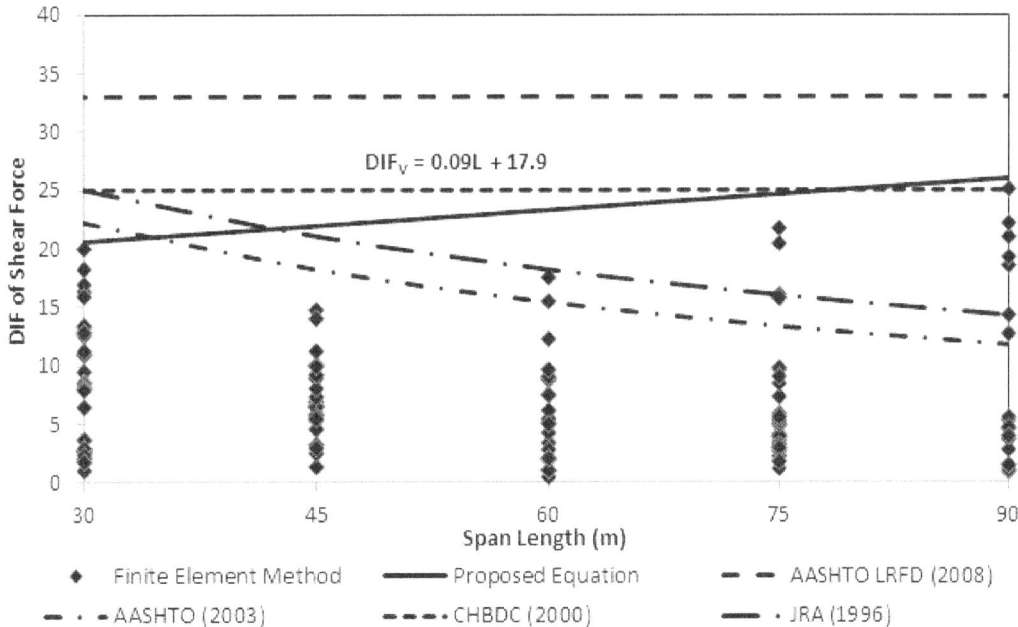

Figure 4.22 Dynamic Impact Factor for Shear Force as Function of Span Length

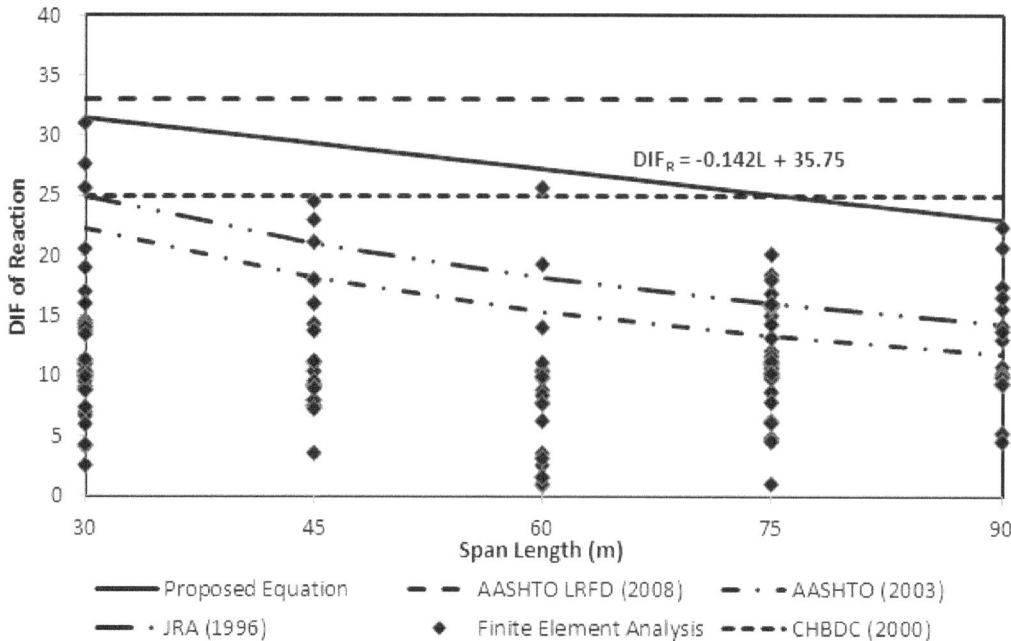

Figure 4.23 Dynamic Impact Factor for Reaction as Function of Span Length

4.6   Summary

The direct integration method was used to determine the dynamic bridge-vehicle interaction results for straight and skew continuous multicell box-girder bridges. Prototype bridge were analyzed using finite element methods subjected to an idealized truck loading condition in all bridge loading lanes. The vehicle speed of 120 Km/h was selected for passing vehicle over both straight and skew bridges which is higher than maximum allowable vehicle speed for safety purposes.

Based on the results of parametric study on prototype bridges, it was observed that the skew angle had significant effect on dynamic impact factor of bridges so that neglecting the skewed bridge in developing of proposed equation for dynamic impact factor lead to unsafe situation. Meanwhile, the number of boxes, number of lane loads and span length did not have a clear relationship with dynamic impact factor of bending moment, shear force and reaction of bridges.

The finite element results were employed to develop proposed equation for dynamic impact factor of bending moment, shear force and reaction. The proposed equations were derived based on the upper bound dynamic impact factor of bridges. The expressions were compared with corresponding results of various current bridge design specifications to determine the preciseness. It was revealed that the AASHTO LRFD (2008) specifications and OHBDC (1983) estimate highly conservative values for dynamic impact factor of bridges, but the CHBDC (2000) obtain sufficiently reasonable value for dynamic impact factor for shear force. Therefore, it was suggested that the proposed equation are used in bridge design procedure to build suitably conservative and economic bridge structures.

CONCLUSIONS

An extensive analytical study was conducted to evaluate the static and dynamic behaviour of continuous straight and skew concrete multicell box-girder bridges. A detailed literature review was performed to set up the basis for this study. The results of the literature review revealed lack of equations to estimate the load distribution factors for this type of bridges. The impact factors expressions available in the AASHTO (2002) standard Specification and AASHTO LRFD (2008) specifications for multicell box-girder bridges were established upon the grillage technique for limited number of actual bridges which does not precisely obtain the complicated nature of actual bridge structure. In addition, there was no extensive study on distribution of torsional moment on superstructure of skew bridges which is needed in defining the behaviour of the bridges subjected to vehicle loading condition. As a result of this lack of knowledge came upon in the literature, the study depicted in this thesis was established with the objective of supplying design engineers and code writers with comprehensively investigated information to improve understanding of static and dynamic behaviour of continuous straight and skew multicell box-girder bridges.

A user-friendly and suitable three-dimensional finite element modeling method was adopted to analyze prototype continuous skew multicell box-girder bridges. In order to verify and substantiate the finite element modeling of the prototype bridges, the results collected from previous field test and analytical studies were compared to results obtained from adopted herein method. The sufficiently well agreement between finding values verified the reliability and accuracy of the results provided from the three-dimensional finite element method.

Three groups of extensive parametric studies were conducted. The first studies were to derive equations for the load distribution factors for tensile stress, compressive stress, and maximum deflection. The effect of the main parameters studied includes: span length, number of lane, number of box, and skew angle were evaluated. Moreover, the effects of other parameters, such as the number of spans, intermediate diaphragms were also investigated. The best-fitting curve technique was used to develop empirical equations for live load distribution factor for stresses and maximum deflection. In order to take into account the effect of solid concrete intermediate diaphragms on live load distribution factor for bending moment and shear force, the results obtained from parametric study were also used to develop two sets of modification factor expressions.

The second parametric study was conducted to deduce skew correction factor expressions for simplified Henry's method. To develop the skew correction factors on the same terms of corresponding LRFD formulas, one hundred of prototypes bridge which design based on Ontario method was investigated. The effect of span-to-depth ratio, skew angle, torsional rigidity and bridge width were considered. The proposed equations were deduced using a nonlinear statistical analysis based on minimum least square fit method. The finding data then used to evaluate the live load distribution of torsional moment, reaction at piers and shear force on bridge superstructure subjected to vehicular loads.

The third parametric study was done to evaluate the dynamic bridge-vehicle interaction. Based on the results obtained from 184 prototype bridges analyzed statically and dynamically under precisely the same vehicle loading conditions, upper bound equations were deduced to estimate the dynamic impact factors for bending moment, shear force and reaction at piers. The empirical equations were derived in two sets as a function of the (a) span length, and (b) fundamental frequency of the bridge. Based on the theoretical studies conducted on continuous skewed concrete cast-in-place multicell box-girder bridges, the following conclusions were drawn:

1. The AASHTO LRFD (2008) specification does not obtain any formula to calculate the live load stress and deflection for skewed multicell box-girder bridges. Therefore, it was necessary to develop the new equations for live load distribution factor live load distribution factors for tensile and compressive stresses, and maximum deflection subjected to the LRFD live loading condition. Using a statistical analysis proposed equations are deduced as a function of skew angle, number of boxes, number of lane loads and span length of bridges. The slightly-greater-than unity average and small standard deviation indicated that the proposed equations determine quit well estimation for the live load distribution factors on multicell box-girder bridges. These factors are not influenced by using solid intermediate diaphragms and increase the number of spans.

2. Due to neglecting the effectiveness of intermediate diaphragms on live load distribution factor of bending moment, the LRFD formulas obtain highly conservative results for both internal and external girders. Three types of intermediate diaphragm (IDs) arrangements includes; bridge with any IDs, parallel to skew support IDs, and perpendicular to girders IDs were considered to understand the effect of intermediate diaphragms on live load distribution factor of bending moment. Based on the collected data from parametric studies, it was concluded that the deck with intermediate diaphragms perpendicular to the longitudinal girder are the best arrangement for load distribution factor in skew bridges. The empirical modification factor equations were deduced to take into account the effectiveness of intermediate diaphragms on live load distribution factor of bending moment from LRFD formulas. The reliability of equations was validated with conducting a comparative study.

3. For the particular bridge considered, the maximum torsional moment is near the obtuse corner of the bridge. It was obtained by putting the truck loads close to the edge of the bridge driven from beginning support to end support. It was also observed that the torsional moment decreased by up to 23% for bridge with skew angle 60°, when torsional stiffness decreased. Diagonal cracking due to exceeding the diagonal principle tension from tensile strength is the main cause of this reduction.

4. Based on the results obtained from parametric study, it was concluded that the dynamic impact factor of bridge do not follow a certain trend when span length, skew angle and number of boxes increased. The observations were completely compatible with results of AASHTO standard specification and other previous study on dynamic impact factor of bridges. The variation of dynamic impact factor versus skew angle indicated that neglecting the skew bridges in developing the expressions for dynamic impact

factor of bridge results in highly unsafe situation for bridge design. Due to complixety of structural behavior of skewed bridges, the dynamic impact factor responses for bending moment, shear force and reaction are different. Thus, two sets of Empirical equations were deduced for dynamic impact factors of shear, moment, reaction and deflection. The equations were established as function of span length in the same term of AASHTO standards specification and CHBDC code, and as function of first fundamental frequency in the same term of OHBDC codes. It was observed that the current specification estimate unreliable values for dynamic impact factor of bridges.

REFERENCES

AASHTO 1993. *American Association for State Highway and Transportation Officials, Guide Specification for Horizontal Highway Bridge, Washington, D.C*

AASHTO 2002. *Standard specifications for highway bridges, American Association of State Highway and Transportation Officials, 17th Ed., AASHTO, Washington, D.C.*

AASHTO 2004. Bridge Design Specifications (2004). *American Association of State Highway and Transportation Officials.*

AASHTO, L. 1996. Standard specifications for highway bridges, American Association of State Highway and Transportation Officials, 16th Edition, Washington, DC.

AASHTO LRFD 2008. *American Association of State Highway and Transportation Officials. Load and Resistance Factor Design (LRFD): Customary US Units. 5th Edition., Washington, D.C*

Akoussah, K. E., Fafard, M., Talbot, M. & Beaulieu, D. 1997. Parametric study of dynamic load amplification factor for simply-supported reinforced concrete bridges. *Canadian Journal of Civil Engineering* 24(2): 313-322.

Al-Rifaie, W. & Evans, H. 1979. An approximate method for the analysis of box girder bridges that are curved in plan. *Proceedings of International Association of Bridges and Structural Engineering, IABSE*: 1-21.

Arizumi, Y., Hamada, S. & Oshiro, T. 1988. Behaviour study of curved composite box girders. *Journal of Structural Engineering* 114(11): 2555-2573.

Ashebo, D. B., Chan, T. H. T. & Yu, L. 2007 a. Evaluation of dynamic loads on a skew box girder continuous bridge Part I: Field test and modal analysis. *Engineering Structures* 29(6): 1052-1063.

Ashebo, D. B., Chan, T. H. T. & Yu, L. 2007 b. Evaluation of dynamic loads on a skew box girder continuous bridge Part II: Parametric study and dynamic load factor. *Engineering Structures* 29(6): 1064-1073.

Aslam, M. & Godden, W. G. 1975. Model studies of multicell curved box-girder bridges. *Journal of the Engineering Mechanics Division* 101(3): 207-222.

AUSTROADS 1992. Bridge Design Code. *Australia's national road authority*. Sydney, Australia.

Bakht, B. & Jaeger, L. G. 1985. *Bridge Analysis Simplified, McGraw-Hill Book Company, New York*

Bakht, B., Jaeger, L. G., Cheung, M. & Mufti, A. A. 1981. The state of the art in analysis of cellular and voided slab bridges. *Canadian Journal of Civil Engineering* 8(3): 376-391.

Beal, D. B. & Kissane, R. J. 1975. Field testing of horizontally curved steel girder bridges, Research Report 27, United States Department of Transportation Washington, D.C.

Billing, J. 1984. Dynamic loading and testing of bridges in Ontario. *Canadian Journal of Civil Engineering* 11(4): 833-843.

Bradford, M. & Wong, T. 1992. Local buckling of composite box girders under negative bending. *Structural Engineer* 70(21): 377-380.

Brady, S. P. & O'Brien, E. J. 2006. Effect of vehicle velocity on the dynamic amplification of two vehicles crossing a simply supported bridge. *Journal of Bridge Engineering* 11(2): 250-256.

Brady, S. P., O'Brien, E. J. & Žnidari, A. 2006. Effect of vehicle velocity on the dynamic amplification of a vehicle crossing a simply supported bridge. *Journal of Bridge Engineering* 11(2): 241–249.

Buchanan, J., Yoo, C. & Heins Jr, C. 1973. Field Study of a Curved Box Beam Bridge, nterim Report No.59, University of Maryland at College Park, Maryland.

Burgueño, R. & Pavlich, B. S. 2008. Evaluation of Prefabricated Composite Steel Box Girder Systems for Rapid Bridge Construction, Department of Civil and Environmental Engineering, Michigan State University, Technical Report.

Cai, C., Chandolu, A. & Araujo, M. 2010. DESIGN-Quantification of Intermediate Diaphragm Effects on Load Distributions of Prestressed Concrete Girder Bridges. *PCI Journal* 54(2): 48-62.

Chandolu, A. 2005. Assessing the Needs for Intermediate Diaphragms in Prestressed Concrete Girder Bridges, Master Thesis, Department of Civil and Environmental Engineering By Anand Chandolu BE, Osmania University.

Chang, D. & Lee, H. 1994. Impact Factors for Simple-Span Highway Girder Bridges. *Journal of Structural Engineering* 120(3): 704-715.

Chapman, J., Dowling, P., Lim, P. & Billington, C. 1971. The structural behaviour of steel and concrete box girder bridges. *Structural Engineer* 49(3): 111-120.

CHBDC 2000. Canadian Highway Bridge Design Code, Canadian Standards Association, Etobicoke, Ontario, Canada.

Cheung, M. 1984. Analysis of continuous curved box-girder bridges by the finite strip method. *In Japanese. Japanese Society of Civil Engineers*: 1-10.

Cheung, M., Bakht, B. & Jaeger, L. G. 1982. Analysis of box-girder bridges by grillage and orthotropic plate methods. *Canadian Journal of Civil Engineering* 9(4): 595-601.

Cheung, M. & Chan, M. 1978. Finite strip evaluation of effective flange width of bridge girders. *Canadian Journal of Civil Engineering* 5(2): 174-185.

Cheung, M. & Cheung, Y. K. 1971. Analysis of curved box girder bridges by finite strip method. *International Association of Bridges and Structural Engineering, IABSE* 31(1): 1-8.

Cheung, M. & Foo, S. 1995. Design of horizontally curved composite box-girder bridges: a simplified approach. *Canadian Journal of Civil Engineering* 22(1): 93-105.

Cheung, M. M. S. & Song, Z. 2009. Finite-Strip Method for the Analysis of Cracked Plates with Application to Plate-Girder Bridges. *Journal of Structural Engineering* 135(2): 198-205.

Cheung, Y. 1968. Finite strip method analysis of elastic slabs. *Journal of the Engineering Mechanics Division, ASCE* 94 EM6: 1365-1378.

Cheung, Y. K. 1969. The analysis of cylindrical orthotropic curved bridge decks. *IABSE Publications* 29(II): 41-52..

Chou, C. C., Uang, C. & Seible, F. 2006. Experimental Evaluation of Compressive Behaviour of Orthotropic Steel Plates for the New San Francisco–Oakland Bay Bridge. *Journal of Bridge Engineering* 11: 140.

Chu, K. H. & Pinjarkar, S. G. 1971. Analysis of horizontally curved box girder bridges. *Journal of the Structural Division* 97(10): 2481-2501.

Chun, B. J. 2010. Skewed bridge behaviours: experimental, analytical, and numerical analysis. Ph.D Thesis, Wayne State University, Detroit, U.S.

Conner, S. & Huo, X. S. 2006. Influence of parapets and aspect ratio on live-load distribution. *Journal of Bridge Engineering* 11(2): 188-196.

CSA 1988. *Canadian Standards Association, Design of highway bridges (CAN/CSA-S6-88), Rexdale, Ontario, Canada.*

CSI 2009. SAP2000-Integrated software for structural analysis and design. Computers and Structures, Berkeley, CA.

Cusens, A. & Loo, Y. 1974. Applications of the finite strip method in the analysis of concrete box bridges. *Proceedings of Institution of Civil Engineers, London.* 57: 251–273.

Dicleli, M. & Erhan, S. 2009. Live Load Distribution Formulas for Single-Span Prestressed Concrete Integral Abutment Bridge Girders. *Journal of Bridge Engineering* 14(6): 472-294.

Ebeido, T. & Kennedy, J. B. 1996. Shear and reaction distributions in continuous skew composite bridges. *Journal of Bridge Engineering* 1(4): 155-173.

Erhan, S & Dicleli, M. 2009. Live load distribution equations for integral bridge substructures. *Engineering Structure* 31(5): 1250-1264.

Euro-code 2 2005. Design of concrete structures-part 1-1: General rules and rules for buildings, European standard, Dublin, Ireland.

Evans, H. R. & Shanmugam, N. E. 1984. Simplified analysis for cellular structures. *Journal of Structural Engineering* 110(3): 531-543.

Fam, A. & Turkstra, C. 1975. A finite element scheme for box bridge analysis. *Computers & Structures* 5(2): 179-186.

Fan, Z. 1999. Field and computational studies of steel trapezoidal box girder bridges. *Ph. D. Thesis, Civil and Environmental Engineering Department*, University of Houston, U.S.

Fan, Z. & Helwig, T. A. 1999. Behaviour of steel box girders with top flange bracing. *Journal of Structural Engineering* 125(8): 829-837.

Fanous, F., May, J.& Wipf, T. 2011. Development of Live-Load Distribution Factors for Glued-Laminated Timber Girder Bridges, *Journal of Bridge Engineering* 16(2): 179–187.

Foinquinos, R., Kuzmanovic, B. & Vargas, L. M. 1997. Influence of diaphragms on live load distribution in straight multiple steel box girder bridges, ASCE. 89-93.

Foinquinos, R., Kuzmanovic, B. & Vargas, L. M. 1997. Influence of diaphragms on live load distribution in straight multiple steel box girder bridges. *Proc, 15th Struct. Congr.*, New York, ASCE. 89-93.

Fu, C. C. & Tang, Y. 2001. Torsional analysis for prestressed concrete multiple cell box. *Journal of Engineering Mechanics* 127(1): 45-51.

Green, T. & Yazdani, N. 2004. Contribution of intermediate diaphragms in enhancing precast bridge girder performance. *Journal of performance of constructed facilities* 18(3): 142-146.

Hambly, E. & Pennells, E. 1975. Grillage analysis applied to cellular bridge decks. *Structural Engineer* 53(7): 267-275.

Hambly, E. C. 1991. *Bridge deck behaviour:* 2nd Edition, E & FN Spon, Taylor & Francis Group, New York.

Hamidi, S. A. & Danshjoo, F. 2010. Determination of impact factor for steel railway bridges considering simultaneous effects of vehicle speed and axle distance to span length ratio, *Structural Engineering* 32(5): 1369-1376.

Hanna, K. E. 2008. Behaviour of adjacent precast prestressed concrete box girder bridges. *Ph. D. Thesis, University of Nebraska, Lincoln, Nebraska, U.S.*

Heins, C. 1978. Box Girder Bridge Design-State of the Art. *American Institute of Steel Construction, Engineering Journal,* 15(4): 126-142.

Heins, C. P. & Lee, W. H. 1981. Curved Box-Girder Bridge: Field Test. *Journal of Structural Division* 107(2): 317-327.

Higgins, C., Turan, O. T., Connor, R. J. & Liu, J. 2011. Unified Approach for LRFD Live Load Moments in Bridge Decks. *Journal of Bridge Engineering* 16(6): 804-811.

Hilber, H. M. & Hughes, T. J. R. 1978. Collocation, dissipation and [overshoot] for time integration schemes in structural dynamics. *Earthquake Engineering & Structural Dynamics* 6(1): 99-117.

Hodson, D. J., Barr, P. J. & Halling, M. W. 2011. Live Load Analysis of Post-Tensioned Box-Girder Bridges. *Journal of Bridge Engineering* 16(6): 777-791.

Hodson, D. J., Barr, P. J. & Halling, M. W. 2012. Live Load Analysis of Post-tensioned Box-Girder Bridges. *Journal of Bridge Engineering* 17(4): 644-651.

Huang, D. 2008. Full-scale test and analysis of a curved steel-box girder bridge. *Journal of Bridge Engineering* 13(5): 492-500.

Huang, H., Shenton, H. W. & Chajes, M. J. 2004. Load distribution for a highly skewed bridge: Testing and analysis. *Journal of Bridge Engineering* 9: 558.

Huo, X., Conner, S. & Iqbal, R. 2003. Re-examination of the Simplified Method (Henry's Method) of Distribution Factors for Live Load Moment and Shear. *Final Report, Tennessee DOT Project No. TNSPR-RES* 1218.

Huo, X., Wasserman, E. & Iqbal, R. 2005. Simplified Method for Calculating Lateral Distribution Factors for Live Load Shear. *Journal of Bridge Engineering* 10: 544.

Huo, X. & Zhang, Q. 2007. Live Load Distribution for Reactions at Piers of Continuous Prestressed Concrete Skewed Bridges. *Proceedings of Structures Congress, ASCE, New Horizons and Better Practices, Reston, Va..*.

Huo, X. S. & Zhang, Q. 2008. Effect of Skewness on the Distribution of Live Load Reaction at Piers of Skewed Continuous Bridges. *Journal of Bridge Engineering* 13(1): 110-114.

Johnston, S. & Mattock, A. 1967. Lateral distribution of load in composite box girder bridges. *Highway Research Record* 167: 25-33.

JRA 1996. Japan Road Association. *Specifications for Highway Bridges, Part 1, Common specifications.* Japan.

Kabir, A. F. & Scordelis, A. C. 1974. Computer Program for Curved Bridges on Flexible Bents. *Structural Engineering and Structural Mechanics Report No. UC/SESM 74-10, University of California, Berkeley, CA.*

Kashif, A. & Humar, J. 1990. Analysis of the Dynamic Characteristics of Box Girder Bridges. *Developments in Short and Medium Span Bridge Engineering. Third International Conference on Short and Medium Span Bridges,* Toronto, Canada. 367-378.

Kashif, A. H. 1992. Dynamic Response of Highway Bridges to Moving Vehicles. *Ph. D Thesis, Department of Civil Engineering, Carleton University,.* Ottawa, Canada.

Khaloo, A. R. & Mirzabozorg, H. 2003. Load Distribution Factors in Simply Supported Skew Bridges. *Journal of Bridge Engineering* 8(4): 241-244.

Kim, Y. J., Tanovic, R. & Wight, R. G. 2013. A parametric study and rating of steel I-girder bridges subjected to military load classification trucks, *Structural Engineering* 56(11): 709-720.

Kozhikote, R. 1989. Analytical and experimental studies on the static and fatigue behaviour of precast prestressed concrete multi-box beam bridge system, Florida Atlantic University. Ph.D Thesis.

Krätzig, W. 1993. 'Best' transverse shearing and stretching shell theory for nonlinear finite element simulations. *Computer Methods in Applied Mechanics and Engineering* 103(1): 135-160.

Kristek, V. 1979. *Theory of box girders*: Wiley-interscience Publication, John Wiley & Sons..

LADOTD 2002. *Bridge Design Manual, Louisiana Department of Transportation and /Development, Baton Rouge, LA.*

Lee, S. Y. & Yhim, S. S. 2005. Dynamic behaviour of long-span box girder bridges subjected to moving loads: Numerical analysis and experimental verification. *International Journal of Solids and Structures* 42(18): 5021-5035.

Li, H. 1992. Thin-walled box beam finite elements for static analysis of curved haunched and skew multicell box girder bridges. *Ph. D Thesis, Department of Civil Engineering, Carleton University, Ottawa, Canada.*

Li, J. & Genmiao Chen, P. 2011. A Method to Compute Live Load Distribution in Bridge Girders. *Practice Periodical on Structural Design and Construction* 16(4): 191-198.

Li, L. & Ma, Z. J. 2010. Effect of Intermediate Diaphragms on Decked Bulb-Tee Bridge System for Accelerated Construction. *Journal of Bridge Engineering* 15(6): 715-722.

Liao, S. S. & Lin, B. H. 1995. Impact formulas for vehicles moving over simple and continuous beams. *Journal of Structural Engineering* 121(11): 1644-1650.

Lim, P., Kilford, J. & Moffatt, K. 1971. Finite element analysis of curved box girder bridges. *Devel Bridge Design and Construction, U.K.:* 264-286.

Lopez-Anido, R. & GangaRao, H. V. S. 1995. Macroapproach Closed-Form Series Solution for Orthotropic Plates. *J. Struct. Eng.,* 121(3): 420-432.

Maisel, B. & Roll, F. 1974. Methods of analysis and design of concrete box beams with side cantilevers. *Technical Report 42.494.* London, Cement and Concrete Association.

Maleki, S. 1991. Compound strip method for box girders and folded plates. *Computers & Structures* 40(3): 527-538.

Mattock, A. H. & Fountain, R. S. 1968. Criteria for design of steel – concrete composite box girder highway bridges. *United States Steel Corporation, Pittsburgh.*

Meyer, C. & Scordelis, A. C. 1971. Analysis of Curved Folded Plate Structures. *AISC Journal of the Structural Division* 97(10): 2459-2480.

Minalu, K. 2010. Finite Element Modelling Of Skew Slab-Girder Bridges, Msc Thesis, Technical University of Delft, Stevinweg, Netherlands.

Moghimi, H. & Ronagh, H. R. 2008. Impact factors for a composite steel bridge using non-linear dynamic simulation. *International Journal of Impact Engineering* 35(11): 1228-1243.

Moon, D. Y., Sim, J. & Oh, H. 2005. Practical crack control during the construction of precast segmental box girder bridges. *Computers & Structures* 83(31): 2584-2593.

O'Connor C & Shaw PA 2000. Bridge Loads: An International Perspective. 1st Edition, SPON Press.

OHBDC 1983. *Ontario highway bridge design code, Ministry of Transportation,.* 2nd Edition, Downsview, Ontario, Canada

Okeil, A. M. & El-Tawil, S. 2004. Warping stresses in curved box girder bridges: case study. *Journal of Bridge Engineering* 9(5): 487-496.

Park, N. H., Choi, Y. J. & Kang, Y. J. 2005. Spacing of intermediate diaphragms in horizontally curved steel box girder bridges. *Finite Elements in Analysis and Design* 41(9): 925-943.

Park, N. H., Lim, N. H. & Kang, Y. J. 2003. A consideration on intermediate diaphragm spacing in steel box girder bridges with a doubly symmetric section. *Engineering Structures* 25(13): 1665-1674.

Patrick, M. D., Huo, X. S., Puckett, J. A., Jablin, M. & Mertz, D. 2006. Sensitivity of live load distribution factors to vehicle spacing. *Journal of Bridge Engineering* 11(1): 131-134.

Razaqpur, A. & Li, H. 1994. Refined analysis of curved thin-walled multicell box girders. *Computers & Structures* 53(1): 131-142.

Razaqpura, A. G., Shedidb, M. & Nofalc, M. 2013. Experimental investigation of load distribution in a composite girder bridge at elastic versus inelastic states, 49(3): 707-718.

Richardson, J. A. & Douglas, B. M. 1993. Results from field testing a curved box girder bridge using simulated earthquake loads. *Earthquake Engineering & Structural Dynamics* 22(10): 905-922.

Saber, A. & Alaywan, W. 2011. Full-Scale Test of Continuity Diaphragms in Skewed Concrete Bridge Girders. *Journal of Bridge Engineering* 16(1): 21-28.

Samaan, M. 2004. Dynamic and ststic analyses of continuous curved composite multiple-box girder bridges. *civil Eng.* Windsor, Ontario, Canada, univ. of Windsor. Ph.D.

Samaan, M., Kennedy, J. B. & Sennah, K. 2007. Impact factors for curved continuous composite multiple-box girder bridges. *Journal of Bridge Engineering* 12(1): 80-88.

Samaan, M. & Sennah, K. 2002. Distribution of wheel loads on continuous steel spread-box girder bridges. *Journal of Bridge Engineering* 7: 175.

Samaan, M., Sennah, K. M. & Kennedy, J. B. 2002. Distribution of wheel loads on continuous steel spread-box girder bridges. *Journal of Bridge Engineering* 7(3): 175-183.

Scordelis, A. 1960. A matrix formulation of the folded plate equations. *ASCE, Journal of the Structural Division* 86: 1-22.

Scordelis, A. C. 1982. Berkeley computer programs for the analysis of concrete box girder bridges. *Proceedings of the NATO Advanced Study Institute on Analysis and Design of Bridges*, Izmir, Turkey, 119-189.

Sennah, K. & Kennedy, J. 1999. Load distribution factors for composite multicell box girder bridges. *Journal of Bridge Engineering* 4(1): 71-78.

Sennah, K. M. 1998. load disstribution factor and dynamic characteristics of curved composite concrete deck-steel cellular bridges, Ph.D. Thesis, University of Windsor, Windsor, Ontario, Canada.

Sennah, K. M., Zhang, X. & Kennedy, J. B. 2004. Impact factors for horizontally curved composite box girder bridges. *Journal of Bridge Engineering* 9(6): 512-520.

Senthilvasan, J., Brameld, G. & Thambiratnam, D. 1997. Bridge–Vehicle Interaction in Curved Box Girder Bridges. *Computer Aided Civil and Infrastructure Engineering* 12(3): 171-182.

Senthilvasan, J., Thambiratnam, D. P. & Brameld, G. H. 2002. Dynamic response of a curved bridge under moving truck load. *Engineering Structures* 24(10): 1283-1293.

Shanmugam, N. & Balendra, T. 1985. Model studies on multi-cell structures. *Proceedings of Institution of Civil Engineers*, London. 79: 55-71.

Shimizu, S. & Yoshida, S. 1991. Reaction allotment of continuous curved box girders. *Thin-Walled Structures* 11(4): 319-341.

Shore, S. & Chaudhuri, S. 1972. Free vibration of horizontally curved beams. *Journal of the Structural Division* 98(3): 793-796.

Siddiqui, A. & Ng, S. 1988. Effect of diaphragms on stress reduction in box girder bridge sections. *Canadian Journal of Civil Engineering* 15(1): 127-135.

Sisodiya, R., Cheung, Y. & Ghali, A. 1970. Finite element analysis of skew, curved box-girder bridge. *International Association for Bridge and Structural Engineering (IABSE)* 30(II):191–199.

Song, S., Chai, Y. & Hida, S. 2003. Live-Load Distribution Factors for Concrete Box-Girder Bridges. *J. Bridge Engrg* 8(5): 273-281.

Tahouni, S. 2003. Design of highway bridge, 2nd Edition. Tehran University Press, Tehran, Iran.

TDOT 1996. Lateral Distribution of Structural Loads, Tennessee Structures Memorandum 043, Tennessee Department of Transportation, Nashville, Tennessee

Théoret, P., Massicotte, B. & Conciatori, D. 2012. Analysis and Design of Straight and Skewed Slab Bridges. *Journal of Bridge Engineering* 17(2): 289-301.

William, K. J. & Scordelis, A. C. 1972. Cellular structures of arbitrary plan geometry. *Journal of the Structural Division* 98(7): 1377-1394.

Wilson, E. L. & Itoh, T. 1983. An eigensolution strategy for large systems. *Computers & Structures* 16(1-4): 259-265.

Yang, D. & Fu, C. C. 1997. Torsional analysis for multiple box cells using softened truss model. *Structural Engineering and Mechanics* 5(1): 21-32.

Yang, M., Qiao, P., Mclean, D. I. & Khaleghi, B. 2010. Effects of overheight truck impacts on intermediate diaphragms in prestressed concrete bridge girders. *PCI Journal* 55(1): 58-78.

Yoo, C., Buchanan, J., Heins, C. & Armstrong, W. 1976. Analysis of a continuous curved box girder bridge. *Transp. Res. Rec., 79, Transportation Research Record*(579): 61-71.

Zhang, H., DesRoches, R., Yang, Z. & Liu, S. 2010. Experimental and analytical studies on a streamlined steel box girder. *Journal of Constructional Steel Research* 66(7): 906-914.

Zhang, Q. 2008. Development of skew correction factors for live load shear and reaction distribution in highway bridge design, Ph.D Thesis, Tennessee Technological University, Tennessee, U.S.

Zhang, X., Sennah, K. & Kennedy, J. 2003. Evaluation of impact factors for composite concrete-steel cellular straight bridges. *Engineering Structures* 25(3): 313-321.

Zheng, L. 2009. Development of new distribution factor equations of live load moment and shear for steel open-box girder bridges, Ph.D Thesis, Tennessee Technological University, Cookeville, Tennessee, U.S.

Zhu, X. & Law, S. 2002. Dynamic load on continuous multi-lane bridge deck from moving vehicles. *Journal of Sound and Vibration* 251(4): 697-716.

Zienkiewicz, O. C. & Taylor, R. L. 2005. *The Finite Element Nethod for Solid and Structural Mechanics*: Butterworth-Heinemann.

Zokaie, T. 2000. AASHTO-LRFD live load distribution specifications. *Journal of Bridge Engineering* 5(2): 131-138.

Zokaie, T., Mish, K. & Imbsen, R. 1993. Distribution of Wheel Loads on Highway Bridges, Phase III. *NCHRP Final Report 12-26 (2), Transportation Research Board, National Research Council, Washington, D.C.*

www.ingramcontent.com/pod-product-compliance
Lightning Source LLC
Chambersburg PA
CBHW081238180526
45171CB00005B/460